Peterson
First Guide
to
Trees

George A. Petrides

Illustrated by
Olivia Petrides
Janet Wehr

HOUGHTON MIFFLIN COMPANY

Boston   New York

To my young granddaughters,
Marisa and Christina Field,
who love the trees in their woods

Copyright © 1993 by Houghton Mifflin Company
Editor's Note copyright © 1993
by Roger Tory Peterson

PETERSON FIRST GUIDES, PETERSON FIELD GUIDES,
and PETERSON FIELD GUIDE SERIES are registered
trademarks of Houghton Mifflin Company.

Library of Congress Cataloging-in-Publication Data

Petrides, George A.
  Peterson first guide to trees / George A. Petrides ; illus-
trated by Olivia Petrides, Janet Wehr.
    p.   cm.
  Includes bibliographical references and index.
  ISBN 0-395-91183-4
  1. Trees—North America—Identification.
  2. Trees—North America—Pictorial works.
  I. Title.
QK110.P45 1993
582.16097—dc20   92-36586
                                              CIP

Printed in Italy
NWI 20 19 18 17 16 15 14 13 12 11

Illustrations and text based on
Field Guide to Eastern Trees
by George A. Petrides, illustrated by Janet Wehr,
and *Field Guide to Western Trees*
by George A. Petrides and Olivia Petrides

Editor's Note

In 1934, my *Field Guide to the Birds* first saw the light of day. This book was designed so that live birds could be readily identified at a distance, by their patterns, shapes, and field marks, without resorting to the technical points specialists use to name species in the hand or in the specimen tray. The book introduced the "Peterson System," as it is now called, a visual system based on patternistic drawings with arrows to pinpoint the key field marks. The system is now used throughout the Peterson Field Guide series, which has grown to nearly 50 volumes on a wide range of subjects, from ferns to fishes, rocks to stars, animal tracks to edible plants.

Even though Peterson Field Guides are intended for the novice as well as the expert, there are still many beginners who would like something simpler to start with—a smaller guide that would give them confidence. It is for this audience—those who perhaps recognize a crow or a robin, buttercup or daisy, but little else—that the Peterson First Guides have been created. They offer a selection of the animals and plants you are most likely to see during your first forays afield. By narrowing the choices—and using the Peterson System—they make identification even simpler. First Guides make it easy to get started in the field, and easy to graduate to the full-fledged Peterson Field Guides. This one gives the beginner a start on the trees of North America.

Roger Tory Peterson

Introducing the Trees

The reasons for learning about trees vary from the purely recreational to the strictly serious. Like nearly all other creatures, people depend totally on green plants. These convert inorganic chemicals into organic foods and also help to maintain essential atmospheric gases in a healthful balance. There is growing concern that people are destroying the environment on which they depend for their prosperity and survival. Many human ills are related to the destruction of plants. The positive effects of green space on morale and property values are clear to almost every city dweller or real estate broker.

A tree, as defined by the U.S. Forest Service, is a woody plant at least 13 feet (4 m) tall with a *single* trunk at last 3 inches (nearly 8 cm) in diameter at breast height ($4^1/_2$ feet; 150 cm). This First Guide covers the more common and widespread trees of the United States and Canada. Heights given are average ranges.

Identifying Unknown Trees

It is easier to make identifications in the field than to collect specimens for later identification at home. In the wild, additional clues are present, such as milky sap, spicy odors, and growth habits. Twig specimens are helpful, along with leaves when they are present.

The first step in identifying a tree is an easy one: deciding whether it has needles or wide leaves. Beyond that, things get a bit trickier, but once you know the difference between simple and compound leaves and can identify a tree's arrangement of leaves as opposite or alternate, you're nearly there.

Leaf Types And Patterns

Plants whose leaves are obviously not needle-like or scale-like are broadleaved plants. Among such plants, a **simple leaf** has only a single blade and is joined by its stalk to a twig or branchlet that is woody. A **compound leaf** has several distinct leaf*lets* attached to a midrib that is *not* especially woody. The base

4

of that midrib is attached only weakly to the woody twig. When either a simple leaf or a complete compound leaf falls or is plucked, a distinct **leaf scar** is left behind on the twig or branchlet. That leaf scar usually has a small **bud** present above or within it. Only an indefinite mark and *no* bud show on the midrib when a leaflet is plucked.

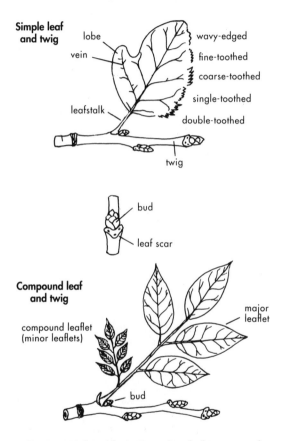

Simple leaf and twig

lobe

vein

leafstalk

twig

wavy-edged

fine-toothed

coarse-toothed

single-toothed

double-toothed

bud

leaf scar

Compound leaf and twig

compound leaflet (minor leaflets)

major leaflet

bud

Distinguishing between simple leaves and the leaflets of a compound leaf soon becomes second nature. At least for a while, however, the novice should always pluck a "leaf" to see

5

whether a leaf scar of definite shape is left behind. This precaution will tend to insure that a leaflet is not misidentified as a leaf.

In a few species (see pages 54 and 58), the **major leaflets** of the compound leaf are themselves divided into **minor leaflets**. Such leaves are **twice-compound** and may have 4 to 800 or more minor leaflets. Both compound and simple leaves may vary in shape, size, texture, and other characteristics.

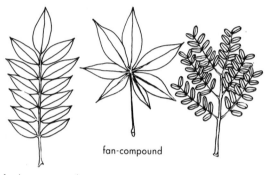

feather-compound

fan-compound

twice-compound

Opposite leaves grow in pairs at opposite sides of the same point on a twig. **Alternate leaves** grow at alternating intervals along the twigs. Some alternate-leaved trees (including apples and cherries) bear small **spur branches** on which leaves and leaf scars are densely clustered. These clusters could be mistaken for opposite foliage if one is not careful to select strong-growing specimen twigs for study.

How to Use This Book

The trees described in this book are classified into six sections, depending on the appearance and arrangement of their leaves. The word **twig** in this book means only the *end* portion of a branch--the newest growth. Any other small branch is called a **branchlet.** The six parts are illustrated on pages 8 and 9. The first step in identifying an unknown plant is to place it in one of the six main groups. Work step by step through the following list, or *key*. Using a key is simply a matter of following a trail that forks repeatedly but usually offers only two paths at each fork. Keys work by listing a series of either-or descriptions that narrow down all possibilities to one, in this case one group of trees. Choose the first description that most closely matches the unknown plant, then choose the best description underneath your first choice and turn to the appropriate section of the book to find the tree.

1. Leaves needlelike or scalelike.

   Part 1, p. 10

1. Leaves broad:
   2. Leaves opposite:
      3. Leaves compound.   **Part 2, p. 30**
      3. Leaves simple.   **Part 3, p. 36**
   2. Leaves alternate:
      4. Leaves compound. **Part 4, p. 46**
      4. Leaves simple.   **Part 5, p. 60**
1. Palms, cacti   **Part 6, p. 122**

Within these sections, trees are grouped according to whether thorns are present and whether leaves or leaflets are evergreen, have smooth or jagged edges, etc.

The main characteristics of each group are illustrated on pages 8–9.

When a tree is without foliage, you must either find leaf remains under the specimen (and run the risk of picking up part of another tree) or rely on twig and other leafless characteristics. The most trying time for the identification of non-evergreen trees is usually early spring, when buds have burst but leaves are small and new twigs are soft. For a few weeks then, some plants may be difficult to identify.

Good specimens are essential for correct identification. Avoid dwarfed, twisted, or gnarled twigs. Abnormally large sucker shoots may not show typical hairiness characteristics, but otherwise strong, quick-growing twigs should be collected for study. On such twigs, the leaves and leaf scars are large, and all details are more evident.

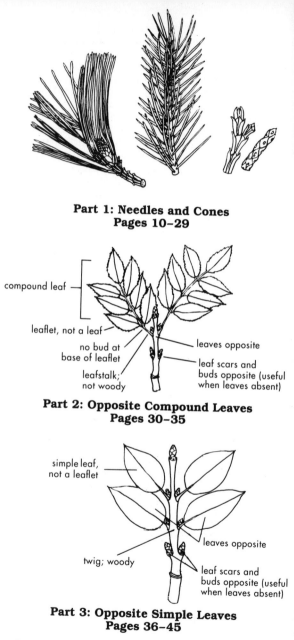

Part 1: Needles and Cones
Pages 10–29

compound leaf

leaflet, not a leaf

no bud at
base of leaflet

leafstalk;
not woody

leaves opposite

leaf scars and
buds opposite (useful
when leaves absent)

Part 2: Opposite Compound Leaves
Pages 30–35

simple leaf,
not a leaflet

leaves opposite

twig; woody

leaf scars and
buds opposite (useful
when leaves absent)

Part 3: Opposite Simple Leaves
Pages 36–45

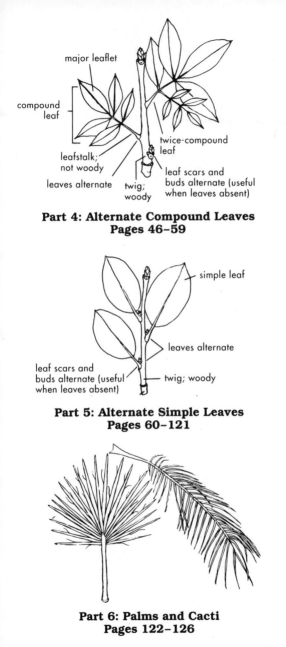

Part 4: Alternate Compound Leaves
Pages 46–59

Part 5: Alternate Simple Leaves
Pages 60–121

Part 6: Palms and Cacti
Pages 122–126

Part 1
TREES WITH NEEDLES AND CONES

This group includes pines, spruces, firs, yews, hemlocks, and others. The needlelike leaves may be long or short, flat or round. Sometimes the needles hug the twigs like scales on a fish. Most species are evergreen, keeping their needles all year. Needleleaf trees are also called conifers because most of them bear fruits called cones. More than 100 conifers are native to North America.

WESTERN PINES

Pines are perhaps our most valuable trees. They provide much of the lumber used in building, and the sap, or pitch, of some species can be made into turpentine.

PONDEROSA PINE   60–130 ft.
Bark on older trees forms flaky yellow plates from which "puzzle pieces" can be removed. The *long* needles grow in clusters of 3; cones are egg-shaped, 3–6 in. long. A common, widespread conifer.

LODGEPOLE PINE   60–100 ft.
A tall, narrow tree of the Pacific Northwest and Rocky Mountains. Needles are sparse and yellow-green, bark is yellowish and scaly. Needles are *short*, growing in clusters of 2. Cones are small. On the coast, it is a much shorter and denser tree (to 30 ft.).

SUGAR PINE   180–200 ft.
This is probably the world's largest pine; certainly it has the *longest cones*, up to 20 in. Needles are blue-green, 2–4 in. long. Grows in mountains of California and Oregon. Sweet pitch oozes from wounds.

BRISTLECONE PINE   15–40 ft.
The longest living tree on earth; the oldest known specimen is at least 4,600 years old. Grows in dry, desert mountains of the Southwest. Needles short, growing in "foxtails."

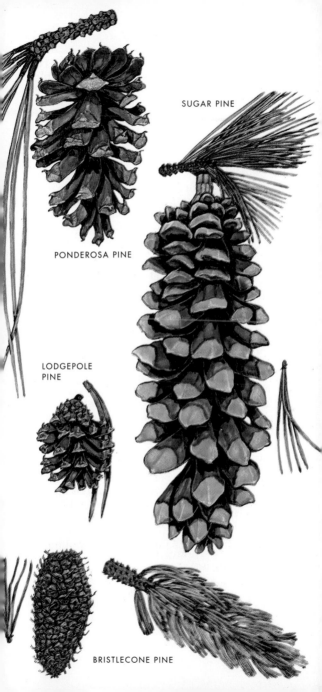

SUGAR PINE

PONDEROSA PINE

LODGEPOLE
PINE

BRISTLECONE PINE

EASTERN PINES

EASTERN WHITE PINE   80–110 ft.
A tall upland tree with the few large limbs growing in horizontal whorls like the spokes of a wheel. Needles are *thin*, 2–4 in. long, in bundles of 5. Cones are slender, 3–10 in. long. The wood is light, soft, and straight-grained.

VIRGINIA PINE   30–40 ft.
One of the several trees sold commercially as "yellow pine," with harder and more res-inous wood than that of white pines. This small or medium-sized tree grows in the Appalachian Mountains and nearby areas. Its needles are 2–3 in. long and grow in clusters of 2. Unlike most pines whose smaller branches snap off cleanly, those of this species are tough and difficult to break. Cones are somewhat egg-shaped, 2–3 in. long; cone scales are tipped with tiny thorns.

JACK PINE   15–40 ft.
An often scrubby 2-needle tree of the North-east and Canada. Needles are very short, only *1–1 1/2 in. long*. The cones, about 2 in. long, are usually curved, *bulging* on one side. Jack Pine produces poor timber, but it is widespread in areas of dry, infertile soil that would otherwise be treeless. Fires cause the cones to release their seeds.

RED PINE   50–80 ft.
A tall straight tree of the Great Lakes region and eastward. The dark green, *4–6 in.* needles are in 2's; the cones are small, roundish, and without thorns. Red Pines are often used in reforestation. Though sometimes called Norway Pine, it is native only to North America.

EASTERN WHITE PINE

VIRGINIA PINE

JACK PINE

RED PINE

13

SOUTHERN PINES

SLASH PINE   80–115 ft.
A handsome species, with needles 5–10 in. long and in clusters of *both* 2's and 3's. (Look at dead needles beneath the tree, if fresh foliage is out of reach.) Buds are *rusty-silver* and cones are *3–6 in.* long. Slash Pine is being planted for timber and paper pulp over ever-widening areas in the South.

LOBLOLLY PINE   80–100 ft.
This pine has needles 6–9 in. long and *3* per cluster. End buds are *brown*. The cones are 3–6 in. long and *prickly* to handle. Loblolly is an important southern timber tree.

SHORTLEAF PINE   90–100 ft.
A tall, mostly southern tree with 3–5 in. needles, usually in clusters of *2*. Cones are small, up to *3 in. long*, and scales have only weak prickles. Shortleaf Pine is used for lumber.

POND PINE   To 80 ft.
A pine of southern lowlands with needles 4–8 in. long and in groups of *3*. Trunk sprouts are often present. The 2–3 in. cones are nearly *round*, prickles weak or absent. The cones often remain attached. Crown branches tend to be more tangled than in other pines.

LONGLEAF PINE   To 85 ft.
A beautiful straight tree with the longest needles and cones of any eastern pine. The needles are in groups of *3* and the cones are *6–10 in. long.* The large *silver-white* end buds are a good field mark. Seedlings are protected from fires by their long grasslike needle clusters. This is a fire-resistant and quick-growing species of value for timber and, at least formerly, for turpentine.

SLASH PINE

LOBLOLLY PINE

SHORTLEAF PINE

POND PINE

LONGLEAF PINE

PINYON PINES

These are pines of the southwestern states that produce delicious *nuts* at the bases of the cone scales. Pinyon nuts formed a principal food of early Native Americans and are still much sought after by residents and visitors alike. They are especially tasty when roasted. Many birds, rodents, and other wildlife also enjoy them.

Pinyon cones are short and round with the scales thick-edged and thornless. Pinyons are members of the white pine group even though they have only 1–4 rather short, stout needles per cluster. There are four pinyon pine species, all trees of arid areas. They are nicely separated by needle number and geographic distribution.

SINGLELEAF PINYON   30–50 ft.
This is one of the easiest pinyons to recognize. It has only one strong, sharp-tipped needle that is nevertheless sheathed at the base like the needle-bundles of other pines.

TWO-NEEDLE PINYON   15–20 ft.
This pine is accurately named. It is widely distributed in Utah, Colorado, and Arizona. It is said to be the most common tree in New Mexico. Resin from trunk wounds is reported to have been used by Native Americans to waterproof woven bottles and to cement turquoise jewelry.

MEXICAN PINYON   15–20 ft.
With 3 needles per cluster, this tree from south of the border reaches the U.S. only in southeastern Arizona, southeastern New Mexico, and extreme west Texas.

PARRY PINYON   15–30 ft.
This could be called 4-needle Pinyon; it is localized in Riverside and San Diego counties in southern California plus a larger area in adjacent Baja California.

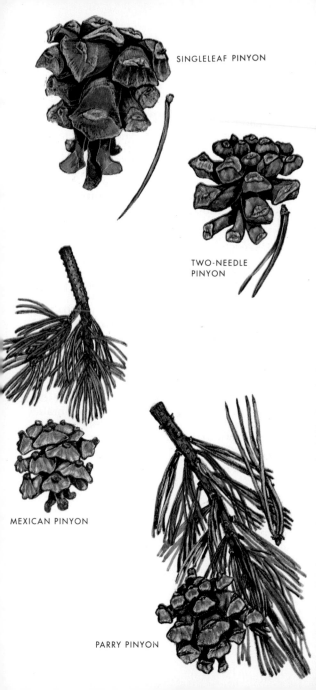

SINGLELEAF PINYON

TWO-NEEDLE PINYON

MEXICAN PINYON

PARRY PINYON

LARCHES

Like pines, these trees also bear slender
needles in groups. Larches, though, have
tufts of flexible needles whorled on *short spur
branches* as well as single needles on longer
shoots. The cones are thin-scaled and *not*
prickly.

Larches are trees of cold latitudes and open
mountain slopes. Their cones are cylindrical
and, unlike most other conifers, the trees turn
golden yellow and *drop their needles* in
autumn.

TAMARACK (AMERICAN LARCH)   40–80 ft.

A tree whose range extends from the arctic
tree line of central Alaska and northern
Canada south to the Canadian prairies and
northeastern U.S. The needles are short and
the cones have *no* papery bracts protruding
beyond the cone scales. Larch lumber is
used for poles, posts, and railroad ties.

WESTERN LARCH   100–180 ft.

This, the largest of our larches, grows
mainly in the Columbia R. basin. The cones
have *pointed* bracts protruding beyond the
cone scales. It is a fire-resistant species protected by bark several inches thick.

SUPALPINE LARCH   30–50 ft.

A tree of the Pacific Northwest mountains
with *white-woolly* twigs. It is often gnarled
and stunted at timberline. The cones have
ragged visible bracts.

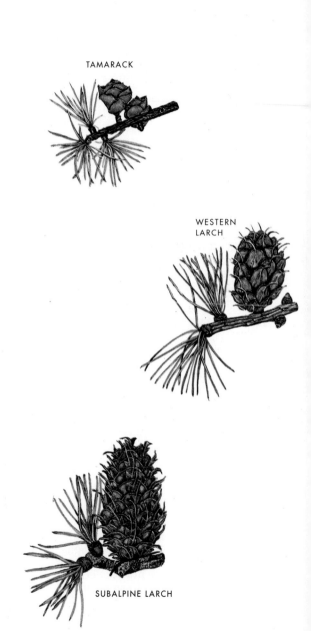

TAMARACK

WESTERN
LARCH

SUBALPINE LARCH

19

SPRUCES

These are narrow-crowned conifers of cold climates whose dead branches are made rough by *tiny wooden pegs*. The needles attached to these pegs are mostly short, stiff, sharp, and 4-sided. The cones are woody and brown, and (unlike fir cones) hang down. Spruces do not make good Christmas trees; their needles dry and fall quickly.

WHITE SPRUCE   50–60 ft.

An important source of paper pulp. Found from the limit of trees on the arctic tundras of Alaska and Canada south to the northeastern U.S. The needles are only $3/8$–$3/4$ in. long, and the cones have *smooth-edged* scales.

BLACK SPRUCE   25–30 ft.

With a distribution similar to White Spruce, this is a smaller tree with shorter needles and *ragged-edged* cone scales. Also valuable for paper pulp.

ENGELMANN SPRUCE   80–120 ft.

The principal spruce of the Rockies and Cascades and a major lumber tree. The needles are dark- to blue-green, over $3/4$ in. long, rather flexible, and *not* very sharp. Cones are mostly $1/2$–2 in. long.

BLUE SPRUCE   80–100 ft.

Much like Engelmann Spruce but growing mainly in the southern Rockies. The needles are *very* sharp and the cones are mostly longer. Widely planted. Wild trees seldom are as blue as cultivated varieties.

SITKA SPRUCE   100–160 ft.

A tree of the foggy northern Pacific Coast with needles *flattened* and white-striped. The twigs *droop* markedly. Certain logs have resonant qualities and are of value for musical instruments. Native Americans wove baskets of the fine rootlets. Our largest spruce, growing tall in some areas but shrubby in far western Alaska.

WHITE SPRUCE

BLACK SPRUCE

ENGELMANN SPRUCE

BLUE SPRUCE

SITKA SPRUCE

FIRS

Like spruces, firs are evergreen trees of cool climates. Unlike spruces, fallen needles of firs leave *smooth, round* leaf scars on the twigs. Also, the short needles are mostly *flat, white-banded* beneath when fresh, and in *flat sprays*. Fir cones are unique: *fleshy*, often purple, and standing *upright*. They *fall apart* early and are mostly gone by autumn.

BALSAM FIR   40–60 ft.

An important pulpwood tree that ranges across Canada to the northeastern U.S. Its resinous knots were once used as torches. Grouse eat the seeds and needles, and snowshoe hares, deer, and moose browse the twigs.

GRAND FIR   150–200 ft.

A large tree of the Pacific Northwest and northern Rockies. The needles are mostly of *2 lengths*. Unlike most conifers, it can grow well even in the shade of other trees. The wood is reported to repel insects. It has an odor that is unpleasant to some people.

WHITE FIR   100–180 ft.

A long-needled fir of the western U.S. mountains. The needles are whitish on both sides; some grow out to the sides, others curve upward to form a *U shape*. Deer browse the twigs and grouse eat the buds and needles. Chipmunks and squirrels eat the seeds.

RED FIR   80–120 ft.

A handsome tree mainly of the Sierras and southern Cascades with needles 4-sided (twirl them between the fingers). The needles are whitened on all surfaces and massed on the twig tops. These trees are found in areas of heavy snow at 5,000–9,000 ft. elevations.

SUBALPINE FIR   20–100 ft.

A common, steeple-shaped fir of the high Rockies and Cascades, shrubby at tree line. Flat needles mostly grow erect, bunching above the twigs.

BALSAM FIR

GRAND FIR

WHITE FIR

RED FIR

SUBALPINE FIR

23

OTHER CONIFERS WITH FLAT NEEDLES

These trees have short needles, mostly on thin *stalks* and in *flat sprays*. Often they are white-banded beneath.

COMMON DOUGLAS-FIR   80–100 ft.

America's most important timber tree and widespread in the West. The twigs hang *drooping* and the cone bracts are uniquely *3-pointed*. Cones are usually common both on the tree and beneath it. Unlike the true firs, Douglas-firs have stalked needles and sharp buds, and their cones are totally different.

EASTERN HEMLOCK   60–70 ft.

The delicate silvery foliage and small, pendent, perfect cones make this northeastern species one of our most beautiful forest trees. The dead twigs of hemlocks are *rough* with needle-pegs but these are much weaker than on the spruces. A tea can be made from the leaves and twigs. Hemlocks do not make good Christmas trees because the needles fall quickly when dry.

REDWOOD   150–325 ft.

Probably the most impressive tree in North America and the world's tallest. Spectacular stands of these wonderful trees occur in the fog belt parks of coastal California. The largest trees are 33 ft. in diameter and 2,000 years old. Despite their large size, they spring from seeds weighing only 1/6000 of an ounce. Trunk is swollen at base, with fibrous bark up to a foot thick. The valuable wood is straight-grained, usually free of knots, and termite resistant.

PACIFIC YEW   25–50 ft.

Formerly considered to be of little economic value, this tree is now much in demand. The bark and needles contain *taxol*, a cancer medicine. This shade-loving species occurs locally in coniferous forests of the Pacific Northwest and northern Rockies. Red, berrylike fruits of yews are unique.

COMMON
DOUGLAS-FIR

EASTERN
HEMLOCK

REDWOOD

PACIFIC YEW

"CEDARS" AND CYPRESSES

Our "cedars" are mostly magnificent trees of
far-western forests. They are not true cedars,
which are native only in the Old World. Their
foliage consists of tiny overlapping leaf *scales*
mostly arranged in *flat sprays*. Like most
conifers, these trees have woody brown cones.

INCENSE-CEDAR 60–80 ft.

A handsome cone-shaped tree of the Northwest mountains. The leaf scales are *shiny*
and in *vase-shaped* whorls. The foliage is
spicy when crushed and does *not* droop. The
cones are *slender, pendent,* and 6-scaled.
The fragrant wood is used in cedar chests.

EASTERN REDCEDAR 40–50 ft.

From the East Coast to the Great Plains,
this juniper tree commonly invades old
fields and pastures. Its spread is aided by
birds which pass the seeds through their
digestive tracts. The aromatic rose-colored
heartwood is used for cedar chests.

WESTERN REDCEDAR 60–130 ft.

A large tree found from Alaska to California
and Montana, whose leaf-scales are *dull* and
whitish beneath. The foliage *droops.* The
urnlike cones are *upright* with 8–12 scales.
The fibrous inner bark was used by Native
Americans for rope and baskets. Canoes
made from the logs could hold 40 people.

ARBOR VITAE 40–50 ft.
(NORTHERN WHITE-CEDAR)

A tree of the Great Lakes region and eastward, with slender $1/2$-in. cones. Supplies
food for moose, deer, squirrels, and birds.
Many varieties are planted as ornamentals.

ARIZONA CYPRESS 40–70 ft.

Ranging from southwestern California to
western Texas, this arid-zone tree has twigs
that *branch at wide angles* and are *not* in
flat sprays. The cones are *round* and often
stay on the tree for months or even years.
Coastal California is home to several other
cypresses.

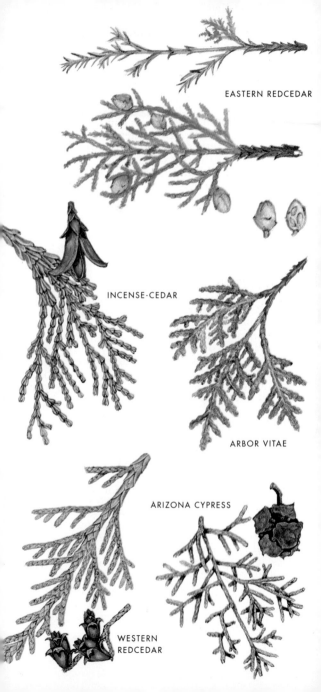

EASTERN REDCEDAR

INCENSE-CEDAR

ARBOR VITAE

ARIZONA CYPRESS

WESTERN REDCEDAR

JUNIPERS AND GIANT SEQUOIA

Junipers resemble cypresses (page 26) in having tiny scalelike leaves. In addition, though, they usually have at least some small, pointed, awl-shaped needles present. Juniper berries are *fleshy*. They are considered to be cones in which the thick scales have become fused. Fruits grow only on female trees. Native Americans and early pioneers ate the raw juniper fruits and made them into a flour. They used the dry shreddy bark of many species for beds and padding. **Eastern Redcedar,** on page 26, also is a juniper.

ROCKY MOUNTAIN JUNIPER　　　15–30 ft.

Growing throughout the Rockies and outlying ranges, this western juniper has *threadlike* scaly twigs only about $1/32$ in. thick. The twigs tend to be 4-sided and usually *droop*.

WESTERN (SIERRA) JUNIPER　　　10–25 ft.

A tree of the dry slopes and high country of the Sierras and Cascades. The twigs are about $1/16$ in. thick.

COMMON JUNIPER　　　To 40 ft.

A northern shrub or small tree, common on old pastures and poor soil. Sharp, whitish needles grow to $7/8$ in. in groups of 3. Fruits are hard, round, blue-black berries.

GIANT SEQUOIA　　　295 ft.

A huge tree of isolated groves in the California Sierras, it may be 50 ft. in diameter with bark up to 2 ft. thick. Some trees have been estimated to be 3,500 years old. Though not reaching the height of some Redwoods (page 24) nor attaining the age of some Bristlecone Pines (page 10), it is the most massive tree in the world and one of the planet's largest organisms (though Australians claim the Great Barrier Reef). One tree, cut down prior to national park establishment, is reported to have required 22 days to fell.

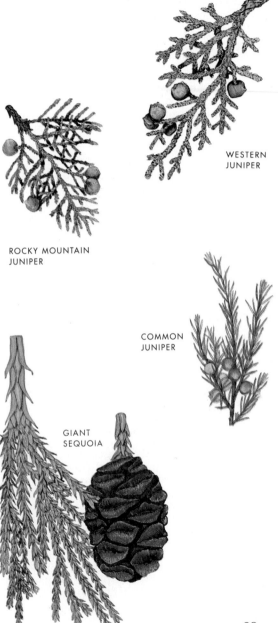

WESTERN
JUNIPER

ROCKY MOUNTAIN
JUNIPER

COMMON
JUNIPER

GIANT
SEQUOIA

Part 2
BROADLEAVED TREES WITH OPPOSITE COMPOUND LEAVES

Relatively few trees have *compound* leaves composed of several leaf*lets*. Even fewer have such leaves in opposing pairs. When the leaves have dropped in winter or during dry seasons, the twigs show opposite leaf scars.

BUCKEYES

Buckeyes are the only native trees with opposite *fan-compound* leaves. The 5–9 toothed leaflets are arranged *like the spokes of a wheel*. The twigs are stout with end buds much larger than the side buds. The flowers are showy and in large, upright, cone-shaped clusters at the twig ends. The blossoms develop into big, brown, shiny nuts that are encased in thick and sometimes prickly husks. Though the seeds are poisonous if eaten raw, Native Americans are reported to have used them for food after grinding them and then passing hot water through the flour.

CALIFORNIA BUCKEYE   To 30 ft.
This is the only wild buckeye on the West Coast. It is found only in California. The May–June display of *pink* flowers is beautiful in some areas of the northern Coast Ranges and Sierra foothills. The end buds are *sticky* and the fruit husks *lack* prickles.

TEXAS BUCKEYE   To 35 ft.
Found from western Missouri to the Edwards Plateau of central Texas. It has 7–11 leaflets per leaf but is sometimes regarded as a variety of the Ohio Buckeye of the East, which has 5 leaflets. Texas and Ohio Buckeyes have *dry* end buds, *yellow* flowers, and fruit husks with *weak* prickles.

HORSECHESTNUT   60–75 ft.
Not a chestnut but a large buckeye imported from Europe and widely planted in parks. It has *sticky* end buds, *white* blossoms, and *thorny* fruit husks.

CALIFORNIA
BUCKEYE

TEXAS
BUCKEYE

HORSECHESTNUT

31

ASHES

Unlike buckeyes (page 30), the ashes have opposite compound leaves with mostly fine-toothed leaflets attached along the midrib in a *feathery* manner. (But note Singleleaf Ash, page 42.) Winter twigs show opposite, shield-shaped leaf scars. The flowers of most ashes are small, dark, without petals, and clustered. The fruits are one-seeded, winged, and shaped like the blades of canoe paddles.

Ashes yield lumber for furniture, tool handles, ladders, baskets, and baseball bats. Twigs are browsed by deer and other animals, and flowers provide pollen for bees. Native Americans once made a dark bitter sugar from the sap of some species. *Fresno* is the Spanish name for ash.

WHITE ASH   70–80 ft.

Our largest and most valuable native ash. The stalked, 5–9 leaflets are pale beneath and may or may not be toothed. Twigs have brown side buds set in a notched leaf scar.

GREEN ASH   85 ft.

An eastern species ranging west to the Great Plains and often planted in windbreaks there. The leaflet stalks are *short* and *narrowly winged*. The foliage may be fine-toothed *or* without teeth. Twigs are smooth and either hairy or not.

TWO-PETAL ASH   To 20 ft.

An unusual ash that has *white flowers*. It ranges through California and south into Baja California Norte. The leaflets have *short* stalks and are sharply *toothed*. The twigs are mostly *4-angled* or 4-lined. The fruits are *winged to the seed base*.

OREGON ASH   To 80 ft.

The only ash found wild in the Pacific Northwest and an important lumber tree. The leaflets are usually *slightly* toothed and *without* stalks. Twigs are hairy and *not* 4-lined. The fruit wing overlaps $^1/_4$–$^1/_2$ the length of the seed.

WHITE ASH

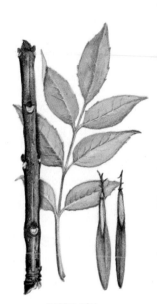

GREEN ASH

TWO-PETAL
ASH

OREGON
ASH

OTHER TREES WITH OPPOSITE COMPOUND LEAVES

Other than buckeyes and ashes, few trees have leaves of this type.

ASHLEAF MAPLE   50–75 ft.

This maple resembles the ashes in having feather-compound leaves. It is widespread except for the Pacific Northwest. Unlike ashes, the 3–5 (7) leaflets are wide with *large, jagged teeth*. Leaf scars are narrow and meet *in raised points* on opposite sides of the green or purplish twigs. Like all maples, the fruits are dry, double-winged "keys," maturing in September–October. The soft white wood is used for boxes. The name Box-elder is widely used, but the tree is not a member of either the box or the elder family. Syrup reportedly can be made from the sap.

ELDERBERRIES

Elderberries may reach small tree size. The twigs are thick but weak, usually with large *pithy* centers. The flowers are small, white, and mostly in *flat-topped* clusters at the twig ends. Fruits are juicy, 3–5-seeded *berries* that can be made into jams, jellies and pies. The berries are eaten by many birds and mammals. Deer and elk browse the twigs. **Common Elderberry,** which grows from 3 to 20 feet tall, is the only elderberry found throughout the eastern U.S. that may reach tree size. It has blackish fruits and white pith. **Blue Elderberry** is widespread in the western mountains. Its fruits are sky blue, dark blue, or black; the pith of its twigs is white or pale brown. **Velvet Elderberry** of California has velvety leaflets and twigs, with white pith. **Mexican Elderberry** of the Southwest has hairless leaflets and brown pith. **Pacific Red Elderberry** occurs along the northern Pacific Coast with bright red fruits in cone-shaped clusters.

ASHLEAF
MAPLE

PACIFIC
RED
ELDERBERRY

BLUE
ELDERBERRY

VELVET
ELDERBERRY

MEXICAN
ELDERBERRY

COMMON
ELDERBERRY

Part 3
BROADLEAVED TREES WITH OPPOSITE SIMPLE LEAVES

Though these trees are more numerous than those of Part 2, they are still few enough to be easily identified. In winter, check the illustrations of Parts 2 and 3 to identify trees with opposite leaf scars.

MAPLES

Maples are our only full-size trees with *opposite fan-lobed* leaves. Most are northern trees. Except for Ashleaf Maple (p. 34), their leaves have 3–5 lobes. Flowers are small, usually greenish, and mostly in short clusters. These develop into paired winged fruits known as "keys." Most maples are of value for shade, ornament, and lumber. Some yield delicious syrup and maple sugar. Maple twigs and seeds are eaten by many animals.

SUGAR MAPLE   80 ft.
One of our most valuable hardwood trees. Both it and the very similar **Black Maple** (not shown) supply sap for maple syrup as well as bird's-eye, curly, tiger, blister, and plain lumber for furniture. The decline of Sugar Maples in some areas is often attributed to acid rain.

RED MAPLE   100 ft.
Though native to the eastern U.S., it is often planted in the West. It has whitish leaf undersides and bright red flowers, twigs, and buds.

SILVER MAPLE   40–60 ft.
An eastern maple whose range extends west into Kansas. Its leaves have large teeth and *deep* indentations between the 5 lobes.

WESTERN MOUNTAIN MAPLE   To 40 ft.
A shrubby tree of western mountains from Alaska to Mexico. Leaves finely toothed. Often called Rocky Mountain Maple or Douglas Maple.

SUGAR
MAPLE

RED MAPLE

SILVER
MAPLE

WESTERN
MOUNTAIN
MAPLE

MANGROVES

Mangroves are small to large trees that grow in shallows along tropical and subtropical coasts. Their root systems slow water movement and provide a place for sand, silt, and debris to collect, thus extending the shoreline seaward. Their leaves are *leathery* and *evergreen*, mostly blunt-tipped, and with the edges rolled under. The fruits germinate and grow larger on the tree before dropping off.

RED MANGROVE   To 80 ft.

A species of southern Florida with *arching prop roots* and often growing in the deeper shallows. The leaves are *green* beneath, usually with tiny black dots. Pale yellow flowers develop into slim, leathery fruits.

BLACK MANGROVE   To 65 ft.

This species has *erect breather roots* and lives in more shallow water. It occurs along the Florida coasts and west to southern Louisiana and southern Texas. The leaves are *white hairy* beneath and the fruits are egg-shaped.

WHITE MANGROVE   To 65 ft.

Much like Black Mangrove, with erect breather roots. Leaves are marked by a pair of *leafstalk glands*. It usually grows landward of both Red and Black mangroves on the coasts of southern Florida.

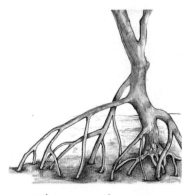

Red Mangrove with prop roots

RED MANGROVE

BLACK
MANGROVE

WHITE
MANGROVE

Black Mangrove
with breather roots

TREES WITH HEART-SHAPED LEAVES

Our only trees with opposite or whorled *heart-shaped* leaves are Princess-tree and Catalpas. In winter, their stout twigs lack central end buds and their circular leaf scars enclose an elliptical ring of tiny bundle scars.

PRINCESS-TREE   30–60 ft.

Also known as Paulownia, this tall Chinese tree has established itself in the wild in the southeastern states. It has large, paired leaves, *chambered* or hollow pith, and large clusters of *purple* spring flowers. The leaves are velvety beneath and *not* whorled. They may have 1 or more large teeth. The *pecan-like* woody fruits contain many small, winged seeds; the husks remain all winter. The trunk bark is rough, interlaced with smooth, often shiny, areas.

CATALPAS

Catalpas are similar to Princess-tree but with *solid* white pith, *white* blossoms, and long *cigar-shaped* fruits. The leaves are paired or in *whorls of 3*. Once with rather restricted ranges, they are now planted widely.

NORTHERN CATALPA   50–70 ft.

A tree of the upper Midwest, this catalpa has long-pointed leaves, flowers 2–3 in. across, and slender fruit pods 10–24 in. long.

SOUTHERN CATALPA   40–50 ft.

Native to the Gulf Coast, this tree has short-pointed leaves. Its flowers are smaller and its fruits wider and shorter than those of Northern Catalpa.

PRINCESS-TREE

NORTHERN
CATALPA

SOUTHERN
CATALPA

41

TREES WITH TOOTHED LEAVES

Only a few trees have opposite, toothed foliage. All are small trees or shrubs.

EASTERN and WESTERN BURNINGBUSHES   6–12 ft.

These small trees have *green* twigs and fleshy reddish fruits beneath woody, 3–5-parted bracts. They are much alike but widely separated geographically.

SINGLELEAF ASH   To 20 ft.

An unusual ash and a resident of the southern Rocky Mountains. It is our only ash with simple leaves (see other ashes on page 32). Occasionally, however, it has 3 leaflets. The leaves (or leaflets, if there are 3) are nearly *circular*. It is common in Zion and Grand Canyon national parks.

NANNYBERRY   9–18 ft.

This is a tree-sized viburnum of the northeastern U.S. and southeastern Canada. It has only 2 bud scales and small, white flowers. The juicy blue-black fruits are 1-seeded and in mostly flat-topped clusters at the twig ends.

RUSTY BLACKHAW   6–18 ft.

A small shrub or tree of southeastern states. Shiny, egg-shaped leaves have *winged leafstalks*. Leafstalks, undersides of leaves, and sometimes twigs have *fine red hairs*. Edible blackish fruits provide food for foxes and birds.

COMMON BUCKTHORN   To 16 ft.

A small European tree or shrub that has become established from Nova Scotia to North Carolina and west to Kansas. Twigs end in fine, pointed spines. Leaves are smooth, with veins curving almost parallel to the leaf edge. Blackish, berrylike fruits are inedible. Other buckthorns (page 102) have alternate leaves.

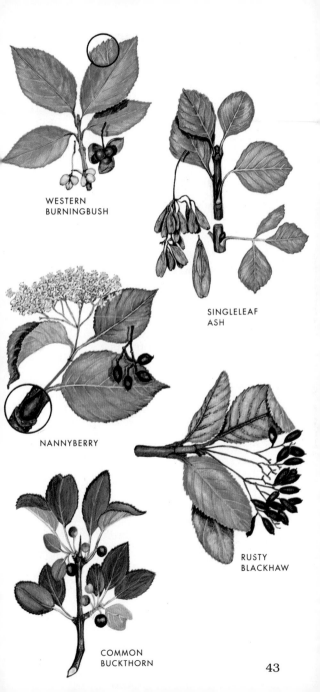

WESTERN
BURNINGBUSH

SINGLELEAF
ASH

NANNYBERRY

RUSTY
BLACKHAW

COMMON
BUCKTHORN

OTHER TREES WITH
OPPOSITE SIMPLE LEAVES

FLOWERING DOGWOOD 10–40 ft.

The blooming of Flowering Dogwood is a
sign of spring in the eastern U.S. The large,
4-petaled, white or pink "flowers," however,
are actually bracts; the true blossoms lie in
small clusters in the center of the bracts.
These blossoms develop into small, fleshy,
red fruits. This is the largest of several,
mostly shrubby, dogwood species. The
trunk bark is dark and *deeply checkered*
like an alligator hide. As in all dogwoods, the
leaf veins *follow* the leaf edges. The similar
Pacific Dogwood ranges in the West Coast
states.

SILVER BUFFALOBERRY 3–7 ft.

The *silver-scaly* leaves, twigs, and buds of
this small tree are distinctive. Found mainly
on the northern plains but scattered else-
where, the twigs are sometimes *thorn-
tipped*. The silver-leaved Russian-olive (page
90) is also often planted as a windbreak or
for ornament but it has *alternate* foliage.

BUTTONBUSH 3–8 ft.

A *wetland* shrub or small tree with leaves
often *whorled* in threes and fours. The veins
do *not* follow the leaf edges. The tiny side
buds are imbedded in the bark. The white
flower clusters and dry brownish fruits are
ball-shaped. Buttonbush is common in California valleys, along the Mexican border
and throughout the eastern U.S.

FRINGETREE 8–18 ft.

Usually a shrub but sometimes tree-size
with beautiful spring flowers. The fruits are
small, juicy, ball-shaped, and purple. It
occurs throughout the southeastern states
where, because of its white drooping blossoms, it is sometimes known as Old Man's
Beard.

FLOWERING
DOGWOOD

SILVER
BUFFALOBERRY

BUTTONBUSH

FRINGETREE

Part 4
BROADLEAVED TREES WITH ALTERNATE COMPOUND LEAVES

Compound leaves have several to many small leaf*lets* attached to a nonwoody midrib. In winter, the alternate leaves of deciduous trees give way to alternate leaf scars.

THORNY TREES WITH ONCE-COMPOUND LEAVES

BLACK LOCUST   70–80 ft.
Native in the East, this species is spreading rapidly in the West and around the world. The 6–20 *rounded* leaflets *lack* teeth. Short *paired* thorns flank buds *hidden* beneath the leaf scars. Fragrant white flower clusters bloom in May–June. Black Locusts make durable fence posts.

DESERT IRONWOOD   25–30 ft.
Growing in Southwestern washes and depressions, this arid-zone tree produces a *dense* mass of grayish, *evergreen* foliage. Lavender blossom clusters are beautiful in May–June. The wood is extremely dense and will not float. Though it dulls tools, Native Americans once made arrowheads and now make bowls and tourist items of it. They also ate the seeds and made bread from the flour.

PRICKLY-ASHES   4–20 ft.
These trees have *toothed* leaflets, *thorny* leafstalks, and a *lemony* odor when crushed. The twigs bear *paired* thorns. The small, greenish flowers yield 1–2 tiny black seeds in small capsules. They are members of the citrus family. Chewing the leaves, twigs, and bark is said to relieve toothache. **Northern Prickly-ash** forms thickets in the Northeast and Midwest. **Southern Prickly-ash** occurs in the Gulf Coast region.

BLACK
LOCUST

DESERT
IRONWOOD

SOUTHERN
PRICKLY-ASH

NORTHERN
PRICKLY-ASH

47

THORNY TREES WITH
TWICE-COMPOUND LEAVES

Here, not only are the leaves divided into leaflets, but these (major) leaflets are subdivided
into smaller (minor) leaflets.

HUISACHE To 30 ft.

An acacia with *paired* thorns, yellow *ball-
shaped* flower clusters, and 8–16 major
leaflets each with 20–50 minor ones. It is
found along the Mexican border and east to
Florida. Deer browse the foliage; bees collect
the pollen. The sap has been used for glue.
Europeans cultivate the flowers for perfume.
Pronounced *weesah-chay*.

GREGG CATCLAW 15–30 ft.

With all too many *single, curved* thorns, this
acacia forms impenetrable thickets from
southern California to southern Texas.
Yellow flower *spikes* produce *twisted* fruit
pods *2–3 in.* long. Native Americans ate the
seeds as a mush. The fruits also are con-
sumed by jackrabbits and quail.

HONEY MESQUITE To 20 ft.

A small tree common on western range-
lands. The *drooping* leaves have only *one
pair* of major leaflets each with 20–40 minor
leaflets. Spur branches form *knobs* between
paired thorns. Once restricted by grass
fires, mesquites grew mainly along stream-
beds. Now with heavy livestock grazing, they
have spread widely. Native Americans made
flour from the seeds and baskets from the
inner bark. Spanish pronunciation is *mes-
keetay*.

HONEY LOCUST 70–80 ft.

Wild trees have *branched* thorns 1–6 in.
long, but a thornless form is planted widely.
The large leaves may be *both* once- and
twice-compound. The fruit pods are *8–18 in.*
long and *twisted*, with sweet pulp between
the seeds. The thorns were once used for
pins, spear points, and animal traps.

HUISACHE

GREGG
CATCLAW

HONEY
MESQUITE

HONEY
LOCUST

49

WALNUTS AND SIMILAR TREES

These trees have leaves that are once-compound and toothed.

BLACK WALNUT   150 ft.

A large eastern tree with 7–17 *spicy-scented* leaflets. The buds are *white-woolly*; cut branchlets show light brown *chambered* pith. The large, hard nuts have thick, 1-piece, stain-producing husks. The dark wood is quite valuable. Five other walnut species range west to California.

BUTTERNUT   40–80 ft.

Like Black Walnut but with darker pith and oblong, 4-lined nut husks. Also called White Walnut for its light-colored wood.

TREE-OF-HEAVEN   80–100 ft.

A Chinese import that grows rapidly under extremely adverse conditions, even in cracks in pavement. Leaflets have only *1–2 pairs* of teeth at base, twigs are thick but weak, and the leaf scars are *large.* The clustered fruits are dry and papery.

MOUNTAIN-ASHES   to 40 ft.

These ornamental trees, unlike true ashes (page 32), have alternate leaves and apple-like fruits. Some grow in arctic areas. Clusters of small, white flowers yield reddish fruits eaten by many birds and mammals. Leaf scars are *narrow.* **American Mountain-ash** and **Showy Mountain-ash** grow in eastern Canada and the northeastern U.S., while the planted **European Mountain-ash** often spreads to the wild there. Western species are similar.

BLACK WALNUT

BUTTERNUT

SHOWY
MOUNTAIN-ASH

EUROPEAN
MOUNTAIN-
ASH

TREE-OF-HEAVEN

AMERICAN MOUNTAIN-ASH

HICKORIES

The hickories are thornless trees with leaves once-compound and toothed. These are aromatic nut trees like walnuts (page 50), but hickories have *solid* branchlet pith and fruits with *4-parted* husks. There are 3 groups: pecans, shagbarks, and pignuts. None of the 11 species ranges west of central Texas. Like walnuts, the crushed green nut husks and foliage are toxic and were used to stun fish, but this is now illegal. Hickory wood is strong but decays rapidly when moist. It was once used for barrel hoops and is an excellent fuel.

PECAN   100–120 ft.

The largest of the hickories, with *9–17* leaflets per leaf and 2–3 pairs of *yellow non-overlapping* bud scales plus *thin, 4-ridged* nut husks. The trunk bark is closely ridged, *not* peeling. Though this hickory originated in the southern Mississippi Valley, about 100 varieties are now cultivated throughout the southeastern states. Nut husks on commercial varieties are thinner than on wild trees.

BITTERNUT HICKORY   50–60 ft.

A wide-ranging tree with smaller leaves than most hickories and 5–11 leaflets. Buds are smoother and brighter yellow than on Pecans.

SHAGBARK HICKORY   60–90 ft.

Also a tall tree, this species has *5–7* large leaflets, stout twigs and end buds *over* $1/2$ in. long with many *overlapping* brownish scales. The trunk bark is *shaggy*, hanging in long, loose strips. The nuts are large, with *thick* husks. They split *to the base*. They are widely collected for food.

PIGNUT HICKORY   80–90 ft.

Though also with 5–7 leaflets, this tree has *slender* twigs (to $1/8$ in. thick), end buds $3/8$–$1/2$ in. long and bark *tight*. The nuts are *under* $1 3/8$ in. long with husks *under* $1/8$ in. thick and *not* splitting to the base.

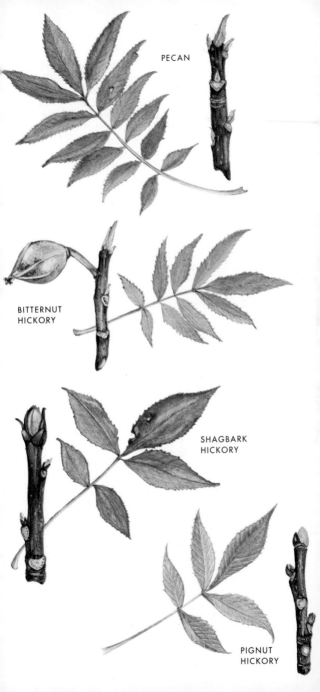

PECAN

BITTERNUT HICKORY

SHAGBARK HICKORY

PIGNUT HICKORY

SUMACS

These thornless trees have once-compound leaves that may or may not be toothed. Most sumacs are small trees or shrubs with tiny, dry, *red-hairy* fruits in upright cone-shaped clusters. These fruits are sometimes made into a "lemonade." They are present most of the year and provide an apparently little-relished but available food supply for wildlife. The twigs are *large*, with hairy buds mostly *hidden* by the leafstalk bases in summer and *almost surrounded* by leaf scars in winter. The pith is brown and solid and the flowers are small and greenish. The bark and leaves have been used to tan leather. The toxic Poison Sumac (not shown) has *white* fruits and is found in swampy areas.

WINGED SUMAC   4–10 ft.
Ranging west to eastern Texas with 11-33 *short-pointed*, shiny leaflets *without* teeth and each about $3/4$ in. wide. Midribs are bordered by conspicuous *"wings"* and the twigs are marked with raised *dots*. The twigs and midribs are often hairy. It is sometimes called Shining Sumac.

PRAIRIE SUMAC   4–10 ft.
Much like Winged Sumac but with leaflets and midrib wings quite *narrow*. The leaflets are *less than* $1/2$ in. wide and *long-pointed*. It grows locally in Texas and adjacent states.

STAGHORN SUMAC   4–15 ft.
This sumac of the northeastern states has 11–31 *toothed* leaflets, and the midribs are *not* winged. The twigs and leafstalks are densely *brown-hairy*, resembling deer antlers in velvet. The twigs are *round* and *not* dotted. Leaf scars are C- or U-shaped.

SMOOTH SUMAC   25 ft.
Found throughout the eastern U.S. and locally in the West. This small tree is similar to Staghorn Sumac, but the twigs and leafstalks are *hairless*, and the twigs are somewhat *flat-sided*.

WINGED SUMAC

PRAIRIE SUMAC

SMOOTH SUMAC

STAGHORN SUMAC

TREES WITH ALTERNATE ONCE-COMPOUND LEAVES

Besides these wild trees, a thornless Honey Locust (page 48) with once-compound leaves is widely planted. None of these trees has toothed leaves.

COMMON HOPTREE   10–20 ft.

The only small tree in the eastern U.S. with alternate *3-parted* leaves. The leaflets are *pointed* and the leaf scars are *U-shaped*. The end leaflet is *short-stalked* and has a narrow V-shaped base. The fruits are flat, papery, and 2-seeded. Reportedly, they have been used as a substitute for hops in flavoring beer. Small specimens could be mistaken for Poison-ivy or Poison-oak, but those plants have end leaflets long-stalked and U-based. California Hoptree has fine-toothed leaves.

TEXAS SOPHORA   18–20 ft.

A *deciduous* tree with 13–15 *rounded* or short-pointed leaflets per leaf. The buds are *hidden* beneath the leaf scars or surrounded by the leafstalk bases. The white flowers are clustered and pealike; the fruits are black beaded pods. Distributed mainly in central and eastern Texas, locally also in nearby Oklahoma, Arkansas, and Louisiana.

MESCALBEAN   TK ft.

An evergreen tree or shrub of central Texas and central Mexico. It has 5–9 shiny, leathery, wide leaflets with blunt tips.

WESTERN SOAPBERRY   20–50 ft.

A deciduous tree of Texas and adjacent states. Its leaves are somewhat leathery, 4–15 in. long, with *8–18* long-pointed leaflets.

BRAZILIAN PEPPERTREE   20 ft.

Introduced from the tropics for its ornamental evergreen foliage and bright red fruits. The species is a pest in southern Florida, however, spreading vigorously to displace native plants. The leafstalks are *red*, and the crushed leaflets give off a turpentine odor.

COMMON HOPTREE

TEXAS SOPHORA

MESCALBEAN

WESTERN SOAPBERRY

BRAZILIAN PEPPERTREE

THORNLESS TREES WITH TWICE-COMPOUND LEAVES

In addition to wild trees of this group, a thornless horticultural variety of Honey Locust (page 48) may have twice-compound leaves. Thorny trees with twice-compound leaves are on page 48.

COFFEETREE   40–60 ft.

Also known as Kentucky Coffeetree, this is an uncommon species native mainly in the Midwest. The leaves are *very large*, being 17–36 in. long. The 5–9 major leaflets bear 8–14 minor leaflets apiece, each such leaflet being 1–3 in. long and *not* toothed. The twigs are *stout*, with *shield-shaped* leaf scars and *sunken* buds. The flower clusters are white and up to 12 in. long. The fruits are brown pods 2–10 in. long with rounded 1 in. seeds. Pioneers made the seeds into a coffeelike drink but the pulp between the seeds is said to be poisonous. This legume does *not* harbor nitrogen-fixing bacteria.

CHINABERRY   To 40 ft.

An oriental import planted widely in the southern states. The leaves are 8–16 in. long with 5–9 major leaflets, each with 3–9 minor ones. The minor leaflets are coarsely *toothed* and the twigs are stout with *3-lobed* leaf scars and *raised* fuzzy buds. The blossoms are purple with an unpleasant odor. The yellowish, globular fruits are *poisonous.* They have been known to paralyze livestock and birds and have been made into a flea powder.

SILKTREE   20–40 ft.

Another Asian native now frequent in the southeastern U.S. The leaves are 5–8 in. long with 6–16 pairs of major leaflets and numerous *smooth-edged* minor ones. When handled, the minor leaflets *close like pages of a book.* The flowers are pink and fluffy; fruits are slim brown pods 2–3 in. long. Sometimes erroneously called mimosa.

COFFEETREE

CHINABERRY

SILKTREE

59

Part 5
BROADLEAVED TREES WITH ALTERNATE SIMPLE LEAVES

THORNY DESERT TREES

To reduce water loss, these trees have tiny leaves present only briefly in spring. The *greenish* twigs and branches function like leaves, using sunlight to make food compounds. *Spine-tipped* twigs lack side thorns.

ALLTHORN   To 15 ft.

Allthorn forms dense thickets in Mexico and in border areas. The twigs are *1–2 in.* long, *stout*, hairless, tipped with *black thorns*, and mostly at *right angles* to the branchlets. The flowers are *greenish*. Quail eat the berries; jackrabbits browse the twigs.

CRUCIFIXION-THORN   15 ft.

A plant of southeastern California and southern Arizona deserts. The *3–5-in.*-long, *stout*, and often *fine-hairy* twigs have *greenish*, scarcely distinguishable, leaf scars. Flower clusters are *reddish-purple*; rings of 5–10 dry, starlike fruits often *remain* for several years. Allthorn may also be called Crucifixion-thorn.

SMOKETHORN   To 25 ft.

A legume of southeastern California and southern Arizona with purple flowers and 1-seeded pea-pod fruits. The *slim, gray-green* twigs are covered with fine whitish hairs, giving the plant a "smoky" appearance.

YELLOW PALOVERDE   To 25 ft.

Found in northwestern Mexico, southern Arizona, and southeastern California. The trunk and branches are *yellow-green* to green, and the 3–6-in. twigs are *moderately stout*, hairless, and with *dark* leaf scars. Has showy yellow flowers in spring. Though a compound-leaved species, it is placed with this group because it is spiny and mostly leafless.

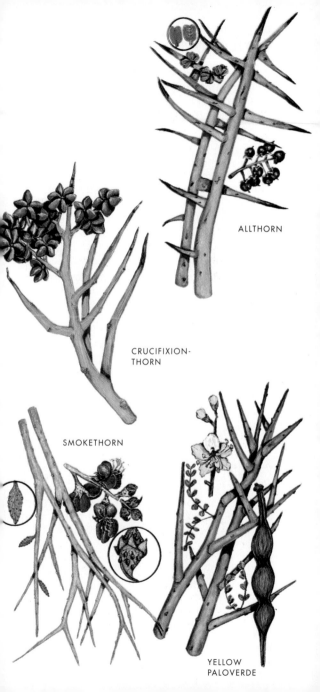

ALLTHORN

CRUCIFIXION-
THORN

SMOKETHORN

YELLOW
PALOVERDE

WILD PLUMS: THORNY TREES WITH TOOTHED LEAVES

Unlike the closely related cherries (page 100), these wild plums are thorny trees with toothed alternate leaves (see also the thornless Garden Plum on page 98). Most such plums have *rounded* leaf teeth and small *glands* on the leafstalks or leaf bases. The distinctive almond odor of broken twigs is less prominent in plums than in cherries. Plum flowers are white or pink, in short clusters, produced from lateral buds or spur branches. The small, round fruits are mostly red or yellow with *one*, usually *flattened*, seed. The trunk bark is usually marked with short horizontal lines.

AMERICAN PLUM   15–30 ft.

An eastern species ranging west to Montana and New Mexico. Its leaves are *dull* and half as wide as long. It is the only thorny plum with *sharp* leaf teeth. The foliage is usually *double-toothed* and *without* glands. The fruits have a *flattened* seed and are eaten by grouse, pheasants, and many songbirds. Some varieties are cultivated.

CHICKASAW PLUM   To 20 ft.

A *narrow-leaved* species forming thickets on the southern plains. It has *shiny*, *single-toothed* and *gland-bearing* foliage. The leaf teeth are *rounded*, the twigs are hairless, and the seeds are *spherical*.

KLAMATH PLUM   To 25 ft.

This plum occurs on rocky slopes below 6,000 feet in elevation in the Coast Ranges and Sierras of Oregon and California. The leaves are egg-shaped to nearly *round*, with the leaf base often heart-shaped. Glands are occasional and the twigs are usually hairless. The fruits are dark red to yellow. Deer browse the twigs and leaves. It is sometimes called Sierra Plum.

AMERICAN PLUM

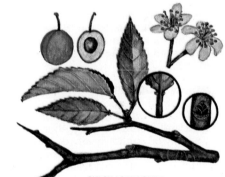

CHICKASAW PLUM

KLAMATH PLUM 63

OTHER THORNY TREES WITH TOOTHED LEAVES

Unlike the thorny plums (page 62), the alternate leaves of hawthorns and crabapples have *no* leafstalk glands and bear fruits with *several* seeds. Hollyleaf Buckthorn has *prickly* foliage.

HAWTHORNS   to 40 ft.

Though distinctive as a group, hawthorns are highly variable and difficult to identify as species. They have *long, slender, smooth* thorns that mostly do *not* bear buds or show leaf scars. They may occur on twigs as well as on older wood. Thornless specimens are seen occasionally. The foliage is toothed and either lobed or not. Fruits are small, yellow to red, and *applelike*, providing food for numerous birds and mammals. Songbirds commonly nest among the thorns. Hawthorns spread as pests into pastures but were planted to make fences in early England.

AMERICAN CRABAPPLE   15–30 ft.

A thicket-forming species of the *north-central* states with leaf bases *round* or heart-shaped and leaf tips mostly *sharp*. Some leaves may be shallowly lobed. The twigs are hairless. The bitter fruits are sometimes used for preserves or vinegar.

NARROWLEAF CRABAPPLE   15–30 ft.

A *southern* species with leaves *wedge-based* and often *blunt-tipped*. The foliage tends to be evergreen in some areas.

HOLLYLEAF BUCKTHORN   To 25 ft.

A *prickly-leaved* evergreen shrub or small tree growing on low mountain slopes in California and Arizona The leaves are only $1/4$–1 in. long, almost round, *leathery*, and with *parallel* side veins. The fruits are small, red, and berrylike.

BLACK
HAWTHORN

NARROWLEAF
CRABAPPLE

AMERICAN
CRABAPPLE

HOLLYLEAF
BUCKTHORN

65

OTHER THORNY TREES

Except for Osage-orange, thorny trees *without* leaf teeth occur mainly in the southern states, both eastern and western, and have leafy or bud-bearing thorns. All usually display spur branches.

GREENBARK CEANOTHUS   20 ft.

Our only thorny tree with simple fan-veined leaves that are not toothed. It occurs in southern California and has *evergreen* and leathery leaves. The leaves have 3 main veins meeting at the U-shaped leaf bases and are blunt-tipped. The twigs tend to alternate at right angles. The fruits are small, dry, black capsules. A component of the chaparral vegetation type and the only member of the large *Ceonothus* genus that is both tree-size and thorny.

GUM BUMELIA   40 ft.

Ranging from Florida to southeastern Arizona and northern Mexico, this small tree has leathery, blunt, V-based foliage with a network of *raised* veinlets beneath. The twigs are *hairy*, often with both spiny tips and side thorns. The twigs form a + when the branch is viewed from the end. The sap is milky and the small fruits are shiny, black and fleshy. Gum collected from freshly cut wood or trunk wounds is some-times chewed. Commercial chewing gum comes from a related tropical species.

OSAGE-ORANGE   60 ft.

Once native to Texas, southeastern Okla-homa and nearby areas, this tree was widely planted for living fences before the invention of barbed wire. It now grows wild in many areas. The leaves are 1–8 in. long, U-based, and somewhat long-pointed. Strong, bare thorns occur at the leaf angles. The sap is milky; the wood is yellow. Fruits are green, fleshy, *grapefruit-sized,* and much *wrinkled.* Because of its reported use by the Osage Indians to make bows, the French name *bois d'arc* (colloquially "bodarc" or "bodock") is still heard.

GREENBARK
CEANOTHUS

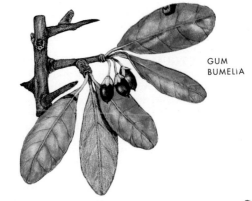

GUM
BUMELIA

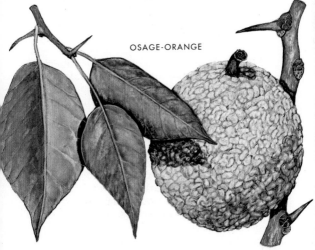

OSAGE-ORANGE

TREES WITH FAN-LOBED LEAVES

Maples (page 36) also have fan-lobed foliage but their leaves are opposite, not alternate.

SYCAMORES   to 175 ft.

These are easy to identify. Their bark is yellowish and flakes off in *puzzlelike* pieces, the leafstalk bases are hollow and cover the buds, the leaf scars surround the buds, the buds have only a single scale, leafy stipules encircle the twigs and leave *rings* on the winter twigs, and fruits hang in tight *balls* on long stems. **Eastern Sycamore** has coarse-toothed leaves and single fruit balls. **California Sycamore** has fine-toothed foliage and 3–7 fruit balls per stalk. **Arizona Sycamore** has leaves not toothed.

SWEETGUM   50–120 ft.

The *star-shaped* and spicy-scented foliage is distinctive, while the ball-shaped fruits are brown and somewhat *prickly*. The wood takes a high polish and is valued for furniture. The gum that exudes from wounds is chewed by some people. A tree of the southeastern U.S.

TULIPTREE   50–100 ft.

A tall and straight eastern tree with unique *4-pointed* leaves, chambered pith, and aromatic buds and foliage. Pairs of leafy stipules enclose the buds and leave encircling rings on the twigs. Large orange and green, tuliplike flowers produce tight clusters of slim, whitish fruits. Native Americans made the trunks into dugout canoes. Though known also as Yellow-poplar and Tulip-poplar, it is a relative of magnolias.

SASSAFRAS   10–50 ft.

This tree has leaves without teeth and in *3 patterns* (3-fingered, mitten, and egg-shaped) on the same tree. The twigs are green and (unlike most trees) often *forked*. All parts of this southeastern tree are pleasantly aromatic; the outer bark of the roots can be boiled to make tea.

CALIFORNIA
SYCAMORE

EASTERN
SYCAMORE

ARIZONA
SYCAMORE

SWEETGUM

TULIPTREE

SASSAFRAS

TREES WITH FAN-VEINED LEAVES

These trees have heart-shaped or triangular foliage with 3 or more main veins meeting at the leaf base.

EASTERN REDBUD   20–40 ft.

A native of the eastern U.S. with leaves heart-shaped, *long-pointed* and *not* toothed. It is planted as an ornamental tree in the West. The red-purple, pealike spring blossoms are sometimes eaten in salads. The roots yield a red dye. **California Redbud** also has heart-shaped leaves but these are short-pointed to *nearly round*. Redbuds are exceptional legume species: unlike other legumes, they do *not* harbor nitrogen-fixing root bacteria.

AMERICAN BASSWOOD   50–80 ft.

The heart-shaped leaves are *toothed* with *uneven* bases. The buds are *green* to bright red with only *2–3* visible scales. Basswood fruits are small nutlets clustered beneath leafy wings that act as spinning parachutes when ripe. When cut, the inner bark of branchlets and roots can be pulled away in fibrous strips and twisted into cords, mats, and lines. Basswood is also called **Linden.**

HACKBERRIES   20–70 ft.

These are trees with *warty* trunks, *rough-hairy* and *uneven-based* foliage, buds with 4–5 scales, and *chambered* pith. The fruits (sometimes called sugarberries) are small, round, 1-seeded, and covered with a thin, sweet, and somewhat edible layer. **Northern Hackberry** is most widespread in the East with leaves *coarsely* toothed. Several other species occur in the South and West, mostly with leaves *finely* toothed or *without* teeth.

EASTERN
REDBUD

AMERICAN
BASSWOOD

NORTHERN
HACKBERRY

71

POPLARS, INCLUDING ASPENS AND COTTONWOODS

Trees with toothed leaves broadly triangular and with 3–5 main veins meeting at the leaf base. The leafstalks mostly are *flattened* near the leaf base, causing the leaves to flutter. Caterpillarlike flower catkins develop into cottony, wind-borne fruits. The trees are fast-growing but short-lived, often colonizing burned or cleared areas. In dry country, they may indicate the presence of underground water near the surface. The seeds, buds, and twigs are eaten by many wildlife species. Beavers consume the inner bark and build dams and lodges with the branches.

QUAKING ASPEN   20–50 ft.

America's most widespread tree, ranging across the continent and from the Arctic to northern Mexico. The *fine-toothed* leaves are nearly *circular*. The buds are *not* gummy. Aspen trunks are white to greenish (but *not* peeling and finely striped like Paper Birch, page 92). Root sprouts are frequent and some vegetative clones may live for centuries. The foliage turns golden yellow in fall.

FREMONT COTTONWOOD   50–75 ft.

This tree, named after a western explorer, ranges along watercourses throughout the southwestern U.S. The triangular, *broad-based*, and often long-pointed leaves are *coarsely* toothed. The buds are *not* gummy. The Spanish name for cottonwood is *alamo*.

EASTERN COTTONWOOD   40–80 ft.

A large tree that ranges throughout the eastern U.S. and northwest to southern Alberta. Its foliage is broadly triangular, *coarse-toothed*, and with 2–3 small *glands* at the leaf base. The buds are *gummy*.

BALSAM POPLAR   30–80 ft.

A large-leaved poplar that ranges from Alaska to Maine. Buds are *gummy* and *spicy-fragrant* when crushed. Leafstalks are mostly *rounded*.

QUAKING
ASPEN

FREMONT
COTTONWOOD

EASTERN
COTTONWOOD

BALSAM
POPLAR

OAKS 1: EASTERN RED OAKS

Not all oaks have lobed leaves (see pages 78–83), but all oaks do have clustered end buds and acorn fruits. Red oaks have hairlike *bristle-tips* on the leaves, and the inner shells (not the cups) of the acorns are *hairy*. These mature over *2* years on the *branchlets*. Most red oaks have dark bark. Both bark and acorns tend to be rich in tannin. Native Americans prepared acorn cakes from flour treated with hot water to remove the tannic acid.

SCARLET OAK   40–50 ft.

This oak is widespread in the East. It is an upland species with leaves deeply lobed, end buds often white-hairy, and *deep* acorn cups. **Pin Oak** is similar but grows mainly on bottomlands with hairless buds, stubby side branches, *shallow* acorn cups, and low branches sloping down.

NORTHERN RED OAK   70–80 ft.

Occurs even in the southern states. It has thin, dull foliage only moderately lobed. The twigs and buds are hairless, and the acorn cup is flat and *saucerlike*. The trunk bark is mostly laced with *shiny strips*.

EASTERN BLACK OAK   70–80 ft.

Like Northern Red Oak in form and range, but the leaves are thicker and *shiny*. The end buds are densely gray-hairy and sharply angled. The acorn cup is *deep*, and the trunk bark is blocky and usually *without* shiny ridges.

SOUTHERN RED OAK   70–80 ft.

Often called Spanish Oak, it has leaves usually with *3 main lobes* toward the tip. The end buds are hairy but *not* angled and the acorn cup is *shallow*. The species ranges north to Long Island and the Ohio Valley.

BLACKJACK OAK   40–50 ft.

A southern oak whose thick, *leathery* leaves are shallowly lobed. The twigs and long buds are *hairy*. its bark is broken into squares.

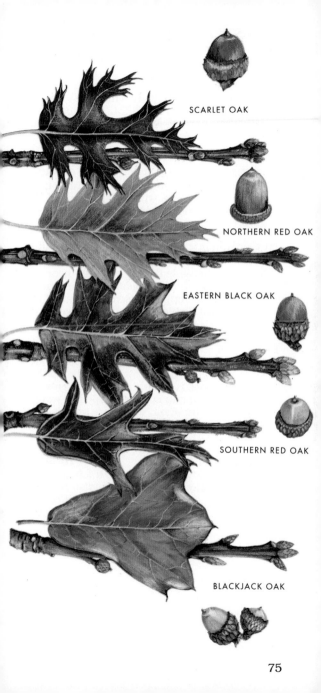

SCARLET OAK

NORTHERN RED OAK

EASTERN BLACK OAK

SOUTHERN RED OAK

BLACKJACK OAK

OAKS 2: EASTERN WHITE OAKS

Trees of the white oak group have *no* bristle-tips on the rounded lobes of their leaves. Their acorns mature the *first year* on the *twigs*, and and the inner shells (not the cups) of the acorns are *hairless*. Other white oaks have *unlobed* leaves (pages 78–83). The trunks are mostly gray-barked. The tannic acid content of bark and acorns is less than that of red oaks.

EASTERN WHITE OAK   60–80 ft.

A tall tree found throughout the East. Its leaves are *hairless*, with 7–11 deep leaf lobes. The buds and twigs also are hairless and the acorn cup is shallow.

OVERCUP OAK   50–80 ft.

Resembles Eastern White Oak but the leaves are usually deeply indented near the base and *fine-hairy* beneath. It is unique in that the acorn cup *encloses* nearly all of the *globular* nut. Principally a tree of swamp forests on the southern coastal plain.

POST OAK   50–60 ft.

Usually a small tree with leaves somewhat *shiny*, *leathery*, and brown-hairy beneath. The 3–5 lobes resemble a *Christian cross* and the acorn cup is bowl-shaped. It ranges throughout the East, with pure stands forming important forest belts in parts of Texas.

BUR (MOSSYCUP) OAK   70–80 ft.

A mainly midwestern tree with foliage usually marked by a *narrow waist*. Slender stipules are frequent among the end buds. The acorn cups have a peculiar "bur" or "mossy" fringe of elongated scales.

EASTERN WHITE OAK

OVERCUP OAK

POST OAK

BUR OAK

OAKS 3: OTHER EASTERN OAKS

Many oaks have leaf edges toothed or wavy-edged or even perfectly smooth. Some are evergreen.

CHINKAPIN OAK   20–50 ft.

Though the foliage resembles that of true chestnuts and chinkapins (page 84), this tree of the Midwest and Appalachia has the acorns and clustered end buds of oaks. The leaves are *sharply* toothed. Like most of the white oak group, the trunk bark is *grayish*. **Chestnut Oak** is similar but with mostly *rounded* teeth and, though a white oak, a *dark, deeply ridged* trunk. It is mainly Appalachian in distribution.

BASKET OAK   To 100 ft.

With leaves wider and often *white-hairy* beneath, this southern oak has *7–16* pairs of *rounded* teeth and sharp end buds. The acorn cup is *short-stalked* and over 1 in. across. The midwestern **Swamp Oak** has similar foliage with only *4–6* pairs of teeth. Its acorns are on *1–3-in.-long* stalks. Both are white oaks and have light gray trunks.

WATER OAK   50–60 ft.

A lowland southern oak whose unlobed, hairless leaves are *widest near the tip* and end in a bristle. The sharp, narrow end buds are *hairy* and *angled*. The acorns have shallow cups, and the dark bark is rather smooth. A member of the red oak groups (see page 74).

VIRGINIA LIVE OAK   To 60 ft.

A spreading southern *evergreen* tree of the white oak group. The leaves are *leathery*, mostly gray- or white-hairy beneath, rarely with a few teeth, and *not* bristle-tipped. The leaf edges are sometimes rolled under. The acorns have stalks $3/4$–1 in. long and the trunk is dark.

CHINKAPIN OAK

BASKET OAK

WATER OAK

VIRGINIA
LIVE OAK

OAKS 4: WESTERN DECIDUOUS OAKS

The several lobed-leaved oaks of the West are all deciduous (non-evergreen). Most are members of the white oak category (see page 76).

CALIFORNIA BLACK OAK   30–75 ft.

Distributed in California and southeastern Oregon, this is the only wild oak of the Pacific slope with *bristle-tipped* leaf lobes. Several eastern trees also of the red oak group, though, may be planted there. California Black Oak has leafstalks *1–2 in.* long and acorns 1–1 1/2 in. long.

OREGON WHITE OAK   40–65 ft.

Another coastal species, extending north to southwestern British Columbia. The leaves are thick, *leathery*, *shiny* above, and *3–5 in.* long. The deep lobes are *not* bristle-tipped, the leafstalks are *1/2–1 in.* long, and the twigs and buds are *hairy*. Acorns are 3/4–1 1/4 in. long.

VALLEY OAK   40–100 ft.

The largest western oak but with leaves only *2–4 in.* long, *thin*, *dull*, and with stalks only *1/4–1/2 in.* long. The foliage and *twigs droop*, and the tree is sometimes called Weeping Oak. The 1–2-in.-long acorns were collected, stored, and eaten in large quantities by early Native Americans. A California species.

GAMBEL OAK   To 65 ft.

The only deeply lobed oak in the southern Rocky Mountains. The foliage is thick, *leathery*, *shiny* and with stalks *1/4–3/4 in.* long. The acorns are 3/4–1 in. long.

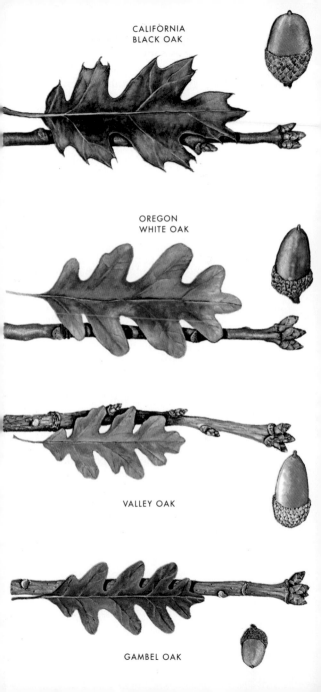

CALIFORNIA
BLACK OAK

OREGON
WHITE OAK

VALLEY OAK

GAMBEL OAK

OAKS 5: WESTERN EVERGREEN OAKS

Evergreen oaks often are called live oaks. Like all oaks, they have leaves and buds clustered at the twig tips. Some have prickly foliage. Coast Live Oak and Emory Oak are sometimes considered to be *intermediate* between red and white oaks, since the acorn shells are hairy inside and yet the nuts mature during their first summer (see pages 74 and 76).

COAST LIVE OAK   30–50 ft.
A broad-topped tree of California's coastal grasslands. The leaves are wide and *arch upward*, umbrellalike. Some leaves may have sparsely prickly edges. The acorns once were much used as food by Native Americans. California quail, mule deer, and other wildlife species also use this food source.

TURBINELLA OAK   15 ft.
This small and often shrubby oak has *flat*, gray-green, prickly, hollylike foliage and acorns with stalks $1/2$–$3/4$ in. long. An arid zone species, ranging across the desert Southwest and south into Mexico. The hybrid with Gambel Oak (page 80) has intermediate characteristics and is often called Wavyleaf Oak.

EMORY OAK   80 ft.
Probably the most abundant oak in the Mexican border region. The *pointed* leaves are shiny on *both* sides, toothed or not, and sometimes prickly. The tree is named after Henry Emory, an early explorer in the Southwest.

SILVERLEAF OAK   60 ft.
Another distinctive oak of the Southwest. The leaves are *narrow*, dark and *shiny* above, *white-* or *silver-hairy* beneath, *pointed* at the tip, and with edges *rolled* under. It is found on mountain slopes from southeastern Arizona to western Texas and northern Mexico.

COAST LIVE OAK

TURBINELLA OAK

EMORY OAK

SILVERLEAF OAK

TREES WITH LARGE LEAF TEETH

These are trees with large, sharp leaf teeth or deeply wavy-edged leaves. All are native to the eastern U.S.

CHESTNUT **To 25 (formerly to 80) ft.**
Formerly a dominant forest tree, but now only a few survivors sprout from old stumps. Soon after 1900, a fungus bark disease from Asia completely eliminated the Chestnut as an important forest tree. Chestnut lumber was quite valuable. The nuts, which occurred *several* to each *spiny* husk, were a staple food of Native Americans and many forms of wildlife. They became marketed widely. It is hoped that blight-resistant strains may be developed.

ALLEGHENY CHINKAPIN **To 40 ft.**
A small tree, similar to Chestnut but with leaves *white-woolly* beneath and nuts only *one* per husk. An Appalachian species but distributed also throughout the Southeast.

BEECH **60–80 ft.**
A distinctive tree with *smooth gray bark and long, slender, many-scaled buds.* The leaves are hairless; leaf teeth are small in some areas. The fruits are small, triangular, edible nuts that are eaten by humans and a wide variety of birds and mammals. The tree grows throughout the East and is widely planted for ornament.

COMMON WITCH-HAZEL **10–25 ft.**
A small tree with *wavy-toothed* and *uneven-based* leaves. The buds are *without* scales, hairy and *stalked* at the base. The spidery, yellow flowers bloom in the autumn *after* the leaves drop and the old fruits pop their seeds up to 20 ft. away. The bark yields an astringent medicine.

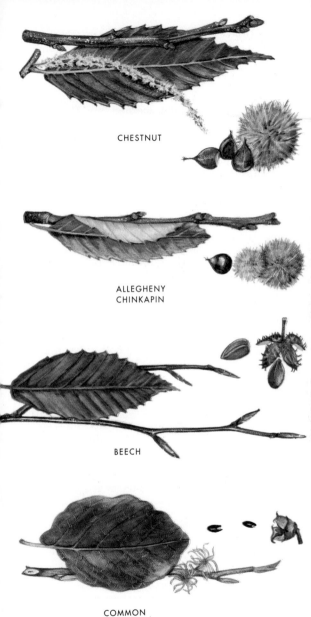

CHESTNUT

ALLEGHENY
CHINKAPIN

BEECH

COMMON
WITCH-HAZEL

WILLOWS 1

Many willows have narrow leaves, but many do not. A better field mark is the *single, smooth bud scale* that characterizes all willows, as well as sycamores (page 68) and magnolias (page 108). Willow leaves are mostly 2–6 in. long, toothed, and the flowers and fruits are in slender catkins. Many willows are only shrubs, some far-north and high-altitude species being only a few inches tall. Willows thrive in damp soil, and help control streambank and mountainside erosion. Willow bark provides salicin, the basic ingredient of aspirin.

SANDBAR WILLOW   To 20 ft.
The most widely distributed willow in North America and the only one with leaves just $1/8$–$3/8$ *in. wide*. The foliage has only a *few* tiny teeth. Native Americans used the twigs in basketry.

BLACK WILLOW   30–100 ft.
Also transcontinental, this is the largest American willow. The leaves are *fine-toothed, shiny*, green on *both* sides, and $1/4$–$5/8$ *in. wide*. The wood is used to make polo balls as well as barrels, doors, and furniture.

PACIFIC WILLOW   To 60 ft.
This willow grows in moist mountain soils from Alaska to southern California and southern New Mexico. Its leaves taper from a U-shaped base to a long, narrow point, with tiny *glands* at the base. Leafstalks are $1/2$–$3/4$ in. long. Once used as a major source for charcoal.

PUSSY WILLOW   To 30 ft.
Ornamental in spring when the furry catkin buds grow large, this willow ranges across Canada and south to Idaho, eastern Tennessee and New Jersey. The leaves are $1/2$–$1 1/2$ *in. wide*, and *coarse-toothed*.

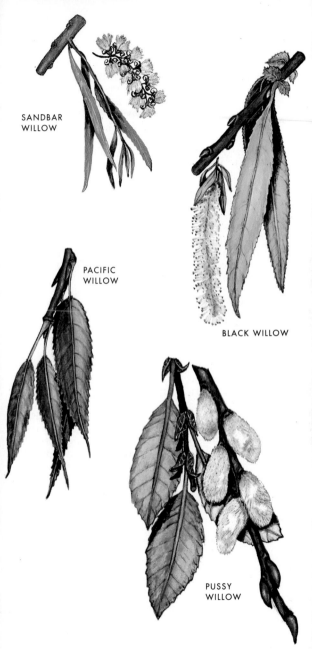

SANDBAR WILLOW

PACIFIC WILLOW

BLACK WILLOW

PUSSY WILLOW

WILLOWS 2

SCOULER WILLOW To 25 ft.

Widespread in the wooded regions of
western North America. The leaves are
mostly *without* teeth, more or less wavy-
edged, and *widest near the* short-pointed or
blunt tip. The twigs often droop. John
Scouler was a Scottish physician who collected plants along the Pacific Coast in the
early 19th century.

BEBB WILLOW To 25 ft.

A common willow in Alaska, Canada, and
the Rocky Mountains. The short leaves are
$1/2$–1 inch wide and taper at both ends, with
fine white or gray hairs on each side. Twigs
also have fine gray hairs. As in some other
willows, the long twigs and branchlets have
been used in basket making.

PEACHLEAF WILLOW To 40 ft.

Found mostly in the Rocky Mountains and
Great Plains, the range of this tree extends
east to the Great Lakes and south to Texas.
The dull, yellow-green leaves are $3/4$–$1^1/4$
in. wide and taper to long points. Twigs are
hairless and tend to droop. Leafstalks are
$1/2$ to $3/4$ in. long.

WEEPING WILLOW 30–50 ft.

An Old World tree easily recognized by its
extremely long twigs and branchlets that
hang vertically, often reaching the ground.
The leaves are relatively narrow, with short
leafstalks. Widely planted.

SCOULER
WILLOW

BEBB
WILLOW

WEEPING
WILLOW

PEACHLEAF
WILLOW

OTHER TREES WITH NARROW LEAVES

Though several trees other than willows have narrow leaves, none of them has the distinctive smooth, one-scale buds of willows.

NARROWLEAF COTTONWOOD   To 60 ft.

Probably no other tree is more willowlike than this one, both in foliage and damp-soil habitat. The slender, fine-toothed foliage, however, has leafstalks *flattened* so that the leaves flutter readily. Also, the buds are *several-scaled*, shiny, and *sticky*. The upper bark is smooth and *whitish*. It occurs mainly along streams in the Rocky Mountains region.

DESERT-WILLOW   To 35 ft.

Ranging across the southwestern states and into northern Mexico, this desert tree has *very* slender, parallel-sided, deciduous leaves. The flowers are large and showy, and the fruits are slim, podlike capsules, 4–12 in. long. It is often planted for ornament.

RUSSIAN-OLIVE   To 25 ft.

Not related to the cultivated olive but frequently planted on the Plains as a windbreak and in towns as an ornamental. The leaves are dark green above and *silvery* beneath. The twigs also are silvery and sometimes thorny. It is often called Oleaster.

CALIFORNIA-BAY   30–80 ft.

An evergreen tree of California and southwestern Oregon. Its shiny, leathery leaves are strongly *spicy-scented* when crushed.

WILLOW OAK   70–80 ft.

A tall, handsome tree with willowlike foliage. The deciduous leaves are *bristle-tipped* and the bark is *dark* like most red oaks. It is widely used in street and park plantings in the South.

NARROWLEAF
COTTONWOOD

DESERT-
WILLOW

RUSSIAN-OLIVE

CALIFORNIA-BAY

WILLOW
OAK

91

BIRCHES

Mostly northern trees with *double-toothed* leaves and *finely cross-striped* trunks. The buds have only *2–3* scales and are brownish and not stalked. The seeds and buds are eaten by grouse and other birds; deer, moose, and other mammals browse the twigs.

PAPER BIRCH   70–80 ft.

An attractive tree with clear white *peeling* bark that separates into layers. (Compare Quaking Aspen, page 72). Native Americans peeled the bark for canoe and wigwam coverings as well as for boxes, cups, makeshift shoes and emergency snow goggles.

WATER BIRCH   To 25 ft.

A western tree found from southwestern Canada south to New Mexico. The bark on its one or more trunks is *shiny, red-brown* with *whitish* cross-stripes, and not peeling. Resembles some cherry trees, but fruits are stout catkins, and broken twigs lack the almond odor of cherries.

SWEET BIRCH   50–70 ft.

A tall brown- or *black-barked* tree mainly of Appalachian and New England forests. The twigs are *hairless* with a spicy *wintergreen* odor when broken. The wood is made into furniture; oil of wintergreen is derived from the sap and leaves. Also called Black Birch.

YELLOW BIRCH   100 ft.

Ranging west to Minnesota and western Ontario, mature trees have shiny *yellow to silver-gray* bark. Young trees resemble Sweet Birch but may have twigs somewhat *hairy* and less strongly aromatic.

RIVER BIRCH   60–80 ft.

This southernmost birch is distinguished by its bark. On young branches, the bark is smooth and *red-brown.* On older branches and trunks, it is *orange, peeling* in untidy curls, and may form rough blackish plates. Buds, twigs, and leafstalks have fine hairs.

PAPER
BIRCH

WATER
BIRCH

SWEET
BIRCH

YELLOW
BIRCH

RIVER
BIRCH

93

ELMS

Elms mostly have *double-toothed* and *uneven-based* leaves. The buds are *many-scaled*, with scales in 2 vertical rows. The fruits are small, flat, oval to *circular*, and *papery-winged*. The inner bark is tough and *fibrous* and can be twisted into fishlines, nets and snares.

AMERICAN ELM   80–100 ft.

Full-sized trees with their divided trunks and vase-shaped form are now rare because of "Dutch" elm disease, a fungus spread by a beetle. The leaves are smooth or rough-hairy above, the buds have *dark-edged* scales, and the fruits are *deeply* notched. It is a tree of eastern lowlands, formerly planted in towns.

SLIPPERY ELM   40–60 ft.

The slimy inner bark of this eastern tree was once well-known as a scurvy preventative. It was ground into flour or chewed piecemeal. The foliage and twigs are quite *sandpapery* while the buds are distinctively *red-hairy*.

WINGED ELM   40–50 ft.

This bottomland elm of the southeastern U.S. usually bears some branches with *wide corky "wings."* The leaves are small and *smooth* above and the bud scales are *dark-bordered*.

SIBERIAN ELM   80 ft.

A hardy shrub or small tree introduced from Asia and widely planted on the Plains as a windbreak. The leaves are only 1–3 in. long, only slightly uneven-based, and mostly *single-toothed*. The buds are *dark* and *blunt*.

SLIPPERY ELM

AMERICAN ELM

WINGED ELM

SIBERIAN ELM

95

OTHER TREES WITH DOUBLE-TOOTHED LEAVES

IRONWOOD   20–40 ft.

A small eastern tree whose smooth *gray* trunk has a *muscular*, rippled look. The lateral leaf veins are *not* forked. The tiny nuts adhere to 3-pointed leafy bracts.

EASTERN HORNBEAM   20–30 ft.

An eastern tree with leaves nearly like those of Ironwood. The trunk bark, however, is *brownish* and *shreddy*, and the leaf veins are mostly *forked*. The fruits are bladder-enclosed nuts. The hornbeams are sometimes known as hophornbeams because the fruits resemble those of hops.

CALIFORNIA HAZELNUT   To 25 ft.

Leaves are nearly circular, *heart-shaped* at the base, and mostly pale and *soft-hairy* (unlike alders) beneath. The edible nuts are often paired and enclosed in a *beaked* husk.

ALDERS

Mainly northern trees that grow in damp soils. The female catkins develop into woody *brown cones* less than 1 in. long. The unusual buds are *reddish* and *stalked*. The leaves and bark are said to control bleeding and diarrhea.

SMOOTH ALDER   6–12 ft.

A more southern version of Speckled Alder, forming thickets next to water. Its leaves are wedge-based, and the trunk is *less spotted.* The woody "cones" do *not* droop.

SPECKLED ALDER   6–12 ft.

Primarily Canadian, this alder spans the continent. The leaves have 9–12 pairs of side veins. The tiny veins on the leaf undersides are parallel, *ladderlike.*

MOUNTAIN ALDER   To 30 ft.

A tree of cold regions from Alaska to the mountains of Arizona and New Mexico. The leaves have only *6–9* pairs of side veins and the veinlets are *irregular.*

IRONWOOD

EASTERN
HORNBEAM

CALIFORNIA
HAZELNUT

SMOOTH
ALDER

SPECKLED
ALDER

MOUNTAIN
ALDER

97

CULTIVATED FRUITS

Unlike American wild cherries (page 100), leaves of Old World cherries are *double-toothed*. The Garden Plum is single-toothed. Most have rounded teeth and *leafstalk glands*. The fruits are fleshy.

SWEET CHERRY   30–50 ft.

Imported from Europe and the parent of many of the sweeter garden cherries. It is a rather tall tree with a *single*, dark main trunk that is prominently marked with horizontal stripes and often peeling. The leaves have *10–14* pairs of side veins and the leaves, flowers, and fruits are clustered on *leafless* spur branches. The fruits have *spherical* seeds. Occasionally the tree escapes to the wild.

GARDEN PLUM   To 25 ft.

This cultivated tree sometimes escapes to grow wild in the Pacific Northwest, northeastern U.S. and southeastern Canada. The leaves are narrow, with *single* teeth. Fruits are purple, up to 1½ inches across and with *flattened* seeds. It is native to eastern Europe and western Asia.

DOMESTIC PEACH   To 25 ft.

Sometimes the peach escapes from plantings to grow wild. The slender foliage is *fine-toothed* and *long-pointed*.

DOMESTIC APPLE   20–30 ft.

A small tree with a *rounded* top and broad, mostly fine-toothed leaves. The twigs and buds are usually somewhat *hairy*. The apple probably originated in the western Himalayas. The apple of the Bible is believed to have been not our northern fruit but the apricot, still common in the Holy Land.

DOMESTIC PEAR   20–35 ft.

Similar to the Domestic Apple but usually with several strong upright branches making a *narrow-topped* tree. The twigs and buds are mostly *hairless*.

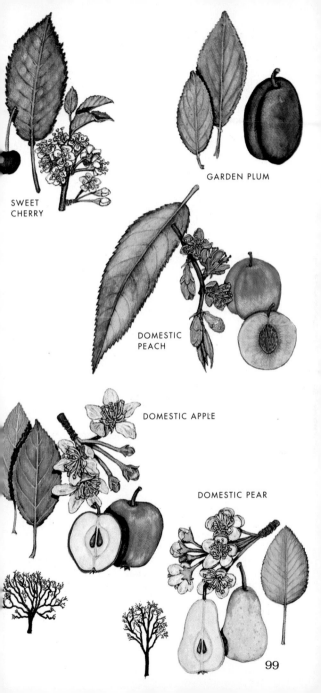

SWEET CHERRY

GARDEN PLUM

DOMESTIC PEACH

DOMESTIC APPLE

DOMESTIC PEAR

99

WILD CHERRIES

Wild cherries have *single-toothed* leaves, leaf-stalks or leafbases with tiny *glands*, broken twigs with a *sour* almond odor, *single* seeds, and trunks with short horizontal *lines*. Wilted foliage may poison livestock. Carolina Laurelcherry (page 116) is evergreen.

CHOKECHERRY   6–20 ft.

Widespread over much of the U.S. and Canada. The leaves are *egg-shaped* with *hairless* midribs and *sharp* teeth. The bud scales have *rounded* tips. Flowers and fruits, unlike most cherries, are in *long, slender clusters*. The tart purplish fruits can be made into jellies and pies. Many song and game birds as well as mammals from rabbits to bears eat the fruit.

BLACK CHERRY   60–80 ft.

Primarily eastern, but ranging into the Southwest and Mexico with *narrow, blunt-toothed* foliage often *hairy-fringed* along the midrib. The bud scales have *pointed* tips. Like Chokecherry, the flowers and fruits are in *slender* clusters. The bitter fruits are eaten by many wildlife species and are often used for jelly. The lumber is highly valued for furniture and house interiors.

FIRE CHERRY   10–30 ft.

Across Canada and the northern U.S., this species invades cleared areas. The leaves are *narrow, hairless, sharp-toothed*, and *crowded* at the twig tips. Flower and fruit clusters are *umbrellalike*. Also called Pin Cherry.

BITTER CHERRY   To 20 ft.

A small tree mainly of the Pacific states. The small, narrow leaves are *round-pointed*. The flowers and fruits are in short, *rounded* clusters. Many birds eat the fruits and mule deer browse the twigs.

CHOKECHERRY

BLACK CHERRY

FIRE CHERRY

BITTER CHERRY

JUNEBERRIES, BUCKTHORNS, etc.

These trees have single-toothed foliage. Juneberry buds are slender and reddish, with
dark-tipped scales; the buds of these buckthorns *lack* scales. Juneberry flowers are in
drooping, white clusters; those of buckthorns
are small and greenish. Juneberries are also
known as serviceberries and shadbushes
(they flower when the shad spawn). The oppo-
site-leaved Common Buckthorn is on page 42.

DOWNY JUNEBERRY   20–40 ft.
Found in the forest understory throughout
the East. The leaves are *fine-toothed, short-
pointed*, heart-shaped at the base, and
somewhat white-hairy beneath. The small
fruits make good jams, jellies and pies.

SASKATOON JUNEBERRY   To 25 ft.
A *coarse-toothed* juneberry that ranges from
Alaska to southern Colorado and western
Minnesota. The leaves often are *nearly circu-
lar*. It is also called Western Juneberry.

CASCARA BUCKTHORN   To 40 ft.
In the Pacific Northwest, the bark of this
tree is harvested commercially for its laxa-
tive qualities. Leaves are wavy-edged or fine-
toothed, with *pointed* tips and 10–15 pairs
of markedly parallel veins *enlarged* beneath.
They may be clustered near the twig tips.
The western buckthorns are not prickly.

BIRCHLEAF BUCKTHORN   To 20 ft.
A southwestern tree with fine-toothed leaves
rounded at the tips and with *7–10 pairs* of
lateral veins. The similar **Carolina Buckthorn,** found further east, has pointed
leaves that smell unpleasant when crushed.

SOURWOOD   20–50 ft.
A tall tree widespread in Appalachia and the
Southeast. Leaves are usually *narrow* and
fine-toothed, sometimes leathery. In June–
July, small, white, bell-like flowers occur in
drooping, twig-end clusters. Dry, 5-parted
fruit capsules may remain until winter. The
leaves turn crimson in autumn.

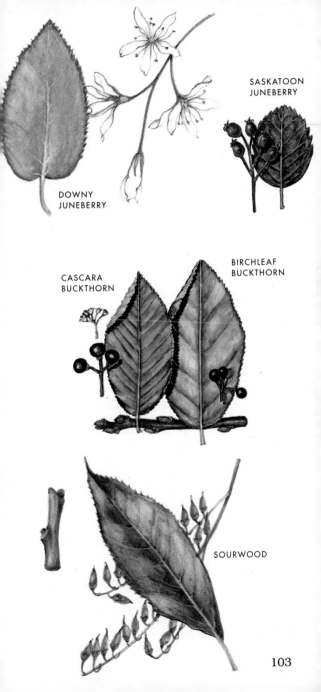

DOWNY
JUNEBERRY

SASKATOON
JUNEBERRY

CASCARA
BUCKTHORN

BIRCHLEAF
BUCKTHORN

SOURWOOD

DECIDUOUS HOLLIES

Only the evergreen American Holly (page 112), the Christmas holly, is easy to identify. The deciduous species have leaves fine- or wavy-toothed and mostly pointed. They are 2–4 in. long, V-based, and often clustered on *spur branches*. Twigs are hairless, buds are mostly pointed, and there may be 2 or more buds at a leaf scar. The flowers are white or greenish and inconspicuous. Holly fruits are small, round, red or orange, and with several seeds.

POSSUMHAW HOLLY　　　　　　　10–20 ft.

An often shrubby species of the southeastern U.S. and Mississippi Valley. The leaves are variable but with *blunt* tips, *narrow* bases, and edges *wavy-toothed*. The side twigs are stiff and the seeds are ridged.

LARGELEAF HOLLY　　　　　　　6–20 ft.

A tall shrub or small tree of the northeastern and mid-Atlantic states. Unlike other deciduous hollies, the leaves are *long-pointed* and *4–6 in. long*. It also has fruits *over 1/2 in.* wide. The side twigs are not especially stiff. The seeds are ridged.

COMMON WINTERBERRY HOLLY　To 25 ft.

A *coarse-toothed* holly widespread in the eastern U.S. and southeastern Canada. The leaves are hairy beneath and the buds are *blunt.* The seeds are smooth. It is sometimes called Black Alder, but this can cause confusion with the true alders (page 96).

CAROLINA HOLLY　　　　　　　To 20 ft.

Mainly a tree of the southeastern coastal plain. The leaves are *fine-toothed* and usually somewhat hairy beneath. The buds are pointed and the seeds are smooth.

POSSUMHAW
HOLLY

LARGELEAF
HOLLY

OMMON
INTERBERRY
OLLY

CAROLINA
HOLLY

BAYBERRIES: LEAVES TOOTHED OR NOT, MAINLY EVERGREEN

These are *spicy-scented* shrubs or small trees with leaves often toothed near the leaf tip. The foliage is mostly marked by yellow resin dots, often too small to see without a magnifying lens. Flower catkins less than 1 in. long grow at the leaf angles in spring. The wax covering of the small, gray, clustered, 1-seeded fruits is sometimes skimmed from hot water and made into scented candles. Often called Waxmyrtle and (with alders) among the few nonlegumes that can enrich the soil with nitrogen through bacterial root nodules.

SOUTHERN BAYBERRY   40 ft.
An *evergreen* tree mainly of the southeastern coastal plain. The foliage is *leathery* and *only 1/2 in. wide*. Resin dots occur on *both* surfaces. The twigs are mostly *hairless*.

EVERGREEN BAYBERRY   15 ft.
Similar to Southern Bayberry but with leaves *1–2 in. wide*, and resin-dotted only *beneath*. The twigs are *black-hairy*.

NORTHERN BAYBERRY   35 ft.
A *non-evergreen* species of the Northeast with thin leaves nearly 1 in. wide. The foliage is often hairy above; resin dots are present mainly on the leaf *underside*. The twigs are *gray-hairy*.

PACIFIC BAYBERRY   35 ft.
Growing wild only near the Pacific Coast from southwestern Washington to southern California. The leaves are *1/2–1 in. wide* with resin dots *beneath*.

BAYBERRIES

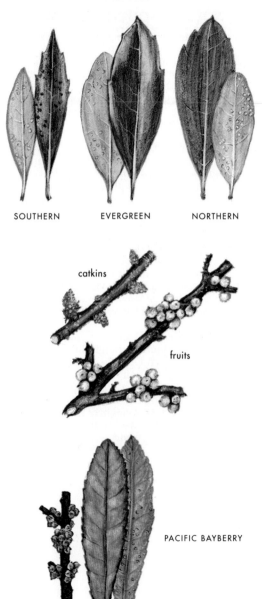

SOUTHERN EVERGREEN NORTHERN

catkins

fruits

PACIFIC BAYBERRY

MAGNOLIAS: SOUTHEASTERN TREES, LEAVES NOT TOOTHED

The magnolias are southern and Appalachian plants with a tropical appearance. The leaves are *smooth-edged* and often large, buds have *one* scale, and the end bud is often *large*. Flowers are mostly big, white, and often fragrant. The conelike fruits release bright red seeds on silky threads.

SWEETBAY MAGNOLIA   To 50 ft.
With foliage and buds *spicy* when crushed, this species ranges north to New Jersey, mostly on the coastal plain. The leaves are somewhat leathery, hairless, and *white beneath*. The buds are *hairy*, the pith is chambered.

SOUTHERN MAGNOLIA   60–80 ft.
This evergreen tree is often planted for its *shiny* foliage, large and fragrant white flowers, and stately size. The buds, twigs, and undersides of leaves are covered with rusty hairs. Pith of the twigs is chambered.

CUCUMBER MAGNOLIA   40–70 ft.
A largely Appalachian magnolia that ranges north to southern Ontario. The thin, pale leaves are slightly hairy beneath. The end bud is *hairy* and *under* $3/4$ in. long. The pith is *not* chambered. The flowers are greenish or yellow and have a slightly unpleasant odor. The dark red seed "cones" look like cucumbers when young.

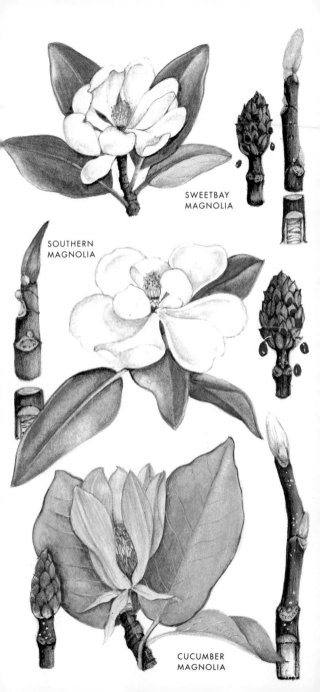

SWEETBAY
MAGNOLIA

SOUTHERN
MAGNOLIA

CUCUMBER
MAGNOLIA

OTHER TREES WITH SMOOTH-EDGED LEAVES

In addition to a few oaks (page 78), willows and willowlike trees (pages 86–91), bayberries (page 106), and magnolias (page 108), these species also have smooth-edged foliage. They range through most of the East.

COMMON PAWPAW   6–20 ft.

This representative of a tropical family has leaves *6–12 in. long.* The end bud is slender, brown-hairy, and *without* scales. Purplish flowers give way to green, somewhat *bananalike*, edible fruits. These are usually eaten by raccoons, opossums, foxes, or squirrels before people find them.

COMMON PERSIMMON   30–50 ft.

An upland tree with the bark broken into small, thick, *dark, squarish blocks.* The leaves are somewhat *thickened* and the small buds are *very dark* and 2-scaled. The pith is usually *solid.* The cherry-sized fruits are eaten by all wildlife, but until they ripen fully, they cause people's mouths seriously to "pucker up." A member of the ebony family.

SOURGUM   40–60 ft.

A lowland tree with *dark, checkered bark* much like that of Persimmon. The buds, however, are *brown* with 4 scales and the pith is *distinctly chambered.* Trees growing in water may have the trunk bases swollen. The small, dark fruits are eaten by bears and many other wild animals. It is also called Black Gum or Black Tupelo.

CRAPEMYRTLE   To 35 ft.

An Asiatic plant of southern towns and abandoned homesites. The twigs are *4-lined* or 4-winged and the buds have 2 scales. The trunk, often with vertical ridges, flakes to a smooth greenish surface. The pink or white flowers are *showy* at twig-ends.

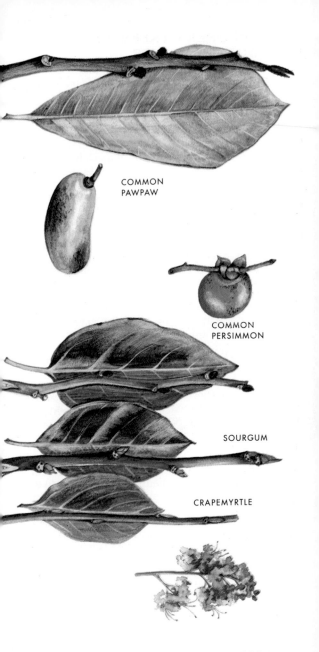

COMMON
PAWPAW

COMMON
PERSIMMON

SOURGUM

CRAPEMYRTLE

EVERGREEN TREES WITH TOOTHED LEAVES

These trees have mostly thick, *leathery* foliage with *coarse teeth*. The leaf edges sometimes are prickly. Except for Loblolly-bay, the fruits are red and juicy.

TOYON　　　　　　　　　　　　To 35 ft.

An attractive tree of California coastal areas, the Sierra foothills, and Baja California. The leaves are sharply toothed but *not* prickly. The white flowers and red fruits are in showy end clusters. It is harvested for decoration and often called Christmasberry, California-holly, and Hollyberry.

AMERICAN HOLLY　　　　　　　10–40 ft.

The universally recognized Christmas holly of the southeastern U.S. The thick, green leaves with *long prickles* are distinctive. The fruits are eaten by numerous songbirds and by quail and wild turkeys. The ivory-white wood is in demand for piano keys, ship models, and inlays.

YAUPON HOLLY　　　　　　　　5–15 ft.

Another distinctive plant of the southeastern U.S. whose red-fruited branches are often gathered for decorations. The leaves are *wavy-edged*, blunt-tipped, and only $1/2$–$1 1/2$ in. long. Native Americans are said to have once brewed a strong medicinal "black drink" from the foliage.

LOBLOLLY-BAY　　　　　　　　To 65 ft.

A tree of the southeastern coastal plain with *large*, thick, dark, shiny leaves and fragrant, white, summertime blossoms 2–3 in. across. The fruits are dry, hairy capsules. A member of the tea family.

TOYON

AMERICAN HOLLY

YAUPON
HOLLY

LOBLOLLY-BAY

113

EVERGREEN TREES WITH FEATHER-TAILED FRUITS

These southwestern trees or small shrubs have leathery leaves and distinctive plumes growing from the end of each fruit. Cercocarpuses are sometimes called mountain-mahoganies because of their dense brown wood, but they are not related to tropical mahoganies.

CLIFFROSE   To 25 ft.

This tree of higher elevations has distinctive *small* leaves (less than $1/2$ in. long) with *3–5 rounded lobes* and edges *rolled under.* The leaves are whitish underneath and slightly sticky. Many *spur branches* are present. The showy spring flowers are creamy or white and about *1 in.* across. The fruits form in groups of 6–10, with plumes *1–2 in. long.*

BIRCHLEAF CERCOCARPUS   To 25 ft.

A tree of the Pacific Coast and central Arizona. The *toothed* leaves are $3/4$–$1 1/2$ in. long with *flat* edges and velvety undersides. The flowers are in *clusters* of 2–5. The fruit plumes are *$1 1/2$–4 in. long.*

CATALINA CERCOCARPUS   To 25 ft.

Only found on California's Santa Catalina Island, this tree resembles Birchleaf Cercocarpus but with leaves *$1 1/2$–3 in. long.* Its flowers and fruits are often *single* rather than clustered.

HAIRY CERCOCARPUS   To 25 ft.

A small-leaved tree of the Mexican border region. The leaf edges are *rolled under* and have a few teeth, and the undersides have fine silky hairs. The fruit plumes are only *1–$1 1/2$ in. long.*

CURLLEAF CERCOCARPUS   To 40 ft.

A widespread shrub or small tree of the western mountains. The smooth-edged leaves are *small, short-stalked, curled-under,* and clustered on spur branches. Flowers are inconspicuous but the fruits have long *(2–3 in.)* feathery tails. The wood is too heavy to float but makes an excellent fuel.

CLIFFROSE

CATALINA
CERCOCARPUS

BIRCHLEAF
CERCOCARPUS

HAIRY
CERCOCARPUS

CURLLEAF
CERCOCARPUS

EVERGREEN TREES WITH SOMETIMES-TOOTHED LEAVES

The foliage of these trees is leathery. Though generally smooth-edged, the leaves occasionally carry teeth.

PACIFIC MADRONE   25–80 ft.

A tree of western coastal districts and the central Sierras with distinctive *smooth red-brown* bark that peels away to show a yellowish layer beneath. The foliage is sometimes *fine-toothed*. The flowers are small, white, and urn-shaped, resembling lilies of the valley. They occur in branched springtime clusters. The fleshy red fruits are eaten by many wildlife species. **Texas Madrone** is similar but is smaller and with leaves less wide. **Arizona Madrone** has narrow leaves and gray, furrowed bark.

SPARKLEBERRY   To 30 ft.

A shrub or small, crooked tree of the South. It is the tallest of the blueberries, though its blackish fruits are less tasty than other blueberries. The elliptical, *short-stalked* leaves are 1–2 in. long with a slender *tip*. Also called Farkleberry.

SUGAR SUMAC   To 15 ft.

One of several evergreen sumacs found on poor soils of southwestern California, Arizona, and northwestern Mexico. The leaves of this species tend to *fold* along the midrib. They have a nice resinous odor and may be sparsely *coarse-toothed*. The flowers are greenish; the fruits are *red-hairy* and reported to be sweet.

CAROLINA LAURELCHERRY   To 40 ft.

A southern evergreen cherry whose *shiny* leaves sometimes bear a *few sharp teeth*. The flowers and fruits are in slender clusters. It is sometimes used in hedges and other ornamental plantings. Wilted foliage, like that of other cherries, may poison livestock.

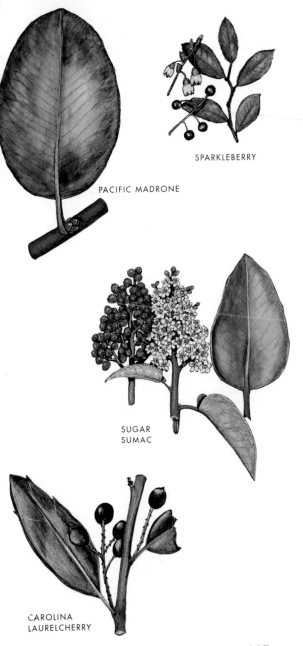

PACIFIC MADRONE

SPARKLEBERRY

SUGAR
SUMAC

CAROLINA
LAURELCHERRY

117

EASTERN EVERGREEN TREES

The leaves of these trees are *leathery* and not toothed. Most (except Redbay) offer beautiful flower displays.

GREAT RHODODENDRON   To 20 ft.

A dense thicket-forming shrub or small tree of slopes and streamsides in the northern Appalachians and New England. The leaves are large and whitish beneath. The edges are rolled under, and both leaf bases and tips are *narrowly pointed*. When the pink-to-purple blossoms are in full bloom during May and June, a rhododendron-covered slope is spectacular.

CATAWBA RHODODENDRON   To 30 ft.

Like the preceding but more southern. The leaves are *rounded* at the base and *broadly* pointed at the tip. Leaf undersides are *light green*.

MOUNTAIN LAUREL   To 10 ft.

A gnarled shrub or small tree that often grows with rhododendrons to form an impenetrable hillside tangle. The leaves are pointed or blunt, *light green* beneath, with *flat* edges. Foliage is often crowded at the twig ends. When an insect lights on the white or purple flowers, one or more stamens springs out of its pocket and slaps the insect with pollen. Thus, pollen is spread to fertilize other plants.

REDBAY   To 50 ft.

A coastal-plain tree whose leaves are narrow, shiny, pale beneath, and *spicy* when crushed. The edges are rolled under and the tips are either pointed or somewhat blunt. The twigs are greenish, hairy, and *angled*.

CYRILLA   To 35 ft.

A southern tree of damp soils and bottom-lands. The shiny leaves are 2–4 in. long with *short* leafstalks and *slightly rounded* tips. White flowers grow at twig bases in clusters 3–8 in. long, becoming small, dry capsules.

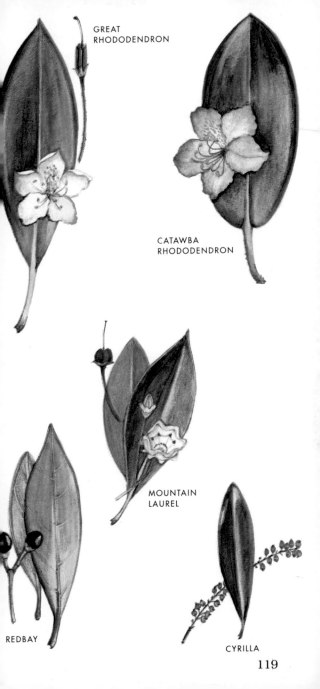

GREAT
RHODODENDRON

CATAWBA
RHODODENDRON

MOUNTAIN
LAUREL

REDBAY

CYRILLA

119

WESTERN EVERGREEN TREES

These are trees with leathery evergreen foliage that is never toothed. These three and their relatives are sparsely distributed in the West.

PACIFIC RHODODENDRON   To 25 ft.

A handsome shrub or small tree mostly of Pacific Northwest coastal areas and Cascades. Large hairless leaves are pointed at *both* ends with the edges *rolled under*. The pink-purple, springtime blossom clusters are very attractive. It is also called California Rosebay.

PARRY MANZANITA   To 15 ft.

One of a large group of mostly shrubby species, of which a few grow to tree size. Manzanitas have *smooth red-brown trunks*, much like those of Pacific Madrone (page 116). The leaves of this species are wide and 1–2 in. long. The springtime flowers are small, pink or white, urn-shaped, and in twig-end clusters.

LAUREL SUMAC   To 15 ft.

Another shrub or small tree, this one found in southwestern California and Baja California, Mexico. The leaves are *pointed* and tend to *fold* along the midrib. The crushed leaves have a slightly unpleasant odor. This plant is one of the first pioneers to return after a fire and is useful in erosion control.

BLUEGUM EUCALYPTUS   70–140 ft.

A quick-growing Australian import that reproduces naturally in the warmer parts of California. The leaves are long, flat, usually *curved*, *long-pointed*, and aromatic when crushed. Leaves of saplings, however, are more round. Falling outer bark and hard, angular fruits make the tree untidy in landscaped areas.

PACIFIC
RHODODENDRON

PARRY MANZANITA

LAUREL SUMAC

BLUEGUM
EUCALYPTUS

Part 6
PALMS AND CACTI

These are tropical and subtropical evergreen trees that are clearly different from all others. The palms have parallel-veined, long-stalked leaves clustered at the ends of the stems. Only a few cacti grow to tree size, and these have uniquely succulent and spiny stems.

FEATHER-LEAVED PALMS

These palms have leaves often 15–20 ft. long. The thornless leafstalks extend the full length of the fronds. Frond segments grow mostly at *right angles* to the midrib. These palms occur mainly in southern Florida, southern California, southern Arizona, and southward.

FLORIDA ROYALPALM   To 125 ft.

The smooth cement-colored, bare, and *bulging* lower trunk topped by a smooth, *bright-green*, cylindrical crownshaft is distinctive. Greenish flowers develop from a spearlike green spathe at the crownshaft base. A tree of rich soils and hammocks (swamp islands).

COCONUT PALM   To 65 ft.

Growing along tropical shores, the wild form of this palm often has a *leaning* trunk with prominent ring scars and a swollen base. A fiberlike *matting* occurs at the bases of the leafstalks. There is *no* crownshaft. The fruits are covered by thick, brown, 3-sided husks. The seeds (coconuts) contain a nutritious "milk" and a lining of tasty white copra ("meat"). Residents of the tropics usually prefer to eat the soft and slippery meat of green (unripe) coconuts.

DATE PALM   To 33 ft.

Cultivated mainly along the Mexican border for the tasty and nutritious fruits. The similar **Canary Island Date Palm** grows to 65 ft. and is planted as a street tree, especially in California.

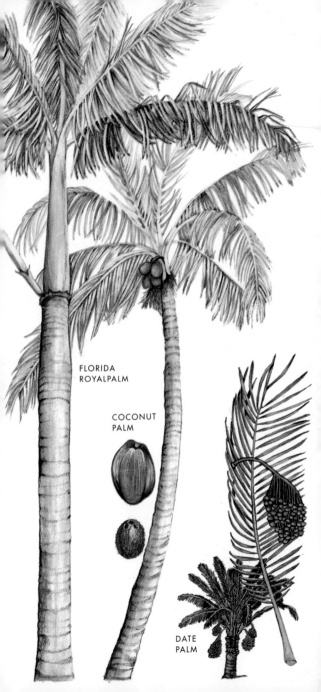

FLORIDA
ROYALPALM

COCONUT
PALM

DATE
PALM

FAN-LEAVED PALMS

These are trees of warm climates with nearly *circular* leaves, the frond segments *radiating* from the end (or from a short "partial midrib") of the leafstalk.

CALIFORNIA WASHINGTONIA   To 50 ft.

A tall palm with leaves 3–6 ft. in diameter, often deeply torn and edged with many *fibrous threads*. The leafstalks are *spiny-toothed*. The trunks may be covered by hanging dead fronds but in landscaped areas these are often removed to reduce dangers of fire and rodent infestation. Native Americans ground the seeds into flour and also ate the growing tips of the tree. As in all palms, this "palm heart" makes a fine salad but its removal kills the tree.

CABBAGE PALM   40–50 ft.

Growing over most of Florida and north along the coast even to southeastern North Carolina, this palm has leaves 4–6 ft. across. It is an unusual fan-palm in that the leafstalk extends *almost completely through* the frond. The leafstalk base is forked and often remains attached to the trunk after the leaf dies.

DWARF PALMETTO   5–15 ft.

Like the Cabbage Palm but grows only to 15 ft. tall and over an even wider southern range. Its leafstalk *barely extends* into the leaf and the base is *not* split.

PAUROTIS PALM   To 25 ft.

An attractive tree of the Florida Everglades. It may have *several* trunks that often *branch*. The leafstalks are edged with stout curved *thorns* and do *not* penetrate the leaf blade. The tree is sometimes used in land-scaping.

PAUROTIS
PALM

DWARF
PALMETTO

CALIFORNIA
WASHINGTONIA

CABBAGE
PALM

125

CACTI

Plants of the desert Southwest with *succulent thorny trunks* and branches.

SAGUARO   To 50 ft.

A widely recognized symbol of the southwestern desert. The *single*, tall, stout, thorny, green trunk is ribbed vertically and *branched* in maturity. The trunk becomes smooth when water is absorbed during rainy periods. The white waxy flowers are about 2 in. across and remain open only about 24 hours. The fleshy red fruits can be eaten fresh or made into preserves. Pronounced *sah-WAHR-oh.*

JUMPING CHOLLA   To 15 ft.

A cactus with *cylindrical* joints covered by inch-long, needle-sharp, barbed thorns. Cholla joints break off to take root. Avoid even brushing against this clumped and impenetrable plant! As one author says, "The Cholla doesn't jump. *You* do." Pronounced *CHOY-yah.*

SAGUARO

JUMPING
CHOLLA

Index

128